Working with Legacy Systems

A practical guide to looking after and maintaining the systems we inherit

Robert Annett

Working with Legacy Systems

Copyright © 2019 Packt Publishing

Author: Robert Annett

Managing Editor: Aditya Shah

Acquisitions Editor: Bridget Neale

Production Editor: Nitesh Thakur

Editorial Board: David Barnes, Mayank Bhardwaj, Ewan Buckingham, Simon Cox, Mahesh Dhyani, Taabish Khan, Manasa Kumar, Alex Mazonowicz, Douglas Paterson, Dominic Pereira, Shiny Poojary, Erol Staveley, Ankita Thakur, and Jonathan Wray

First Published: May 2019

Production Reference: 1310519

ISBN: 978-1-83898-256-0

Published by Packt Publishing Ltd.

Livery Place, 35 Livery Street

Birmingham B3 2PB, UK

Table of Contents

Preface

About

This section briefly introduces the author and the coverage of this book.

About the Course

There comes a point in everyone's IT career when they become responsible for a legacy system. This is inevitable and I call it the 'Penelope Principle.' Like the Peter Principle (where people are promoted to their level of incompetence) and the Dilbert Principle (incompetent workers are promoted to where they can do least damage – management), this should be accepted and worked with rather than being fought against. It is stated thus:

"All IT workers will be promoted into a position where they become responsible for a legacy system" (The caveat is "unless the worker is so useless, they can't hold down a job and keep getting fired before this occurs." This, of course, does not apply to any reader sensible enough to buy this book. (And, yes, I do know this is the 'No True Scotsman' fallacy)).

Hopefully, you are nodding your head vigorously at this point (having bought this book, this is highly likely) and the purpose of this book is to help you deal with the situation you have found yourself in.

Why is this the case? In *Chapter 1, Definition, Issues and Strategy* I will spend a short while defining exactly what we mean by a 'legacy system,' but it comes down to success and longevity. The commercial IT revolution started in the 1970s when vast numbers of manual processes and physical records were placed into mainframe systems. Subsequently, these systems have not only been improved but entire new industries have been created. There are vast benefits to having information systems in electronic form and this effect was magnified by the internet revolution of the 1990s. Some systems don't add value and are scrapped, but most do and are therefore used until it becomes cost-effective to replace them.

This means there are a LOT of IT systems out there involved with every aspect and function of society. This has been happening for almost 50 years, so the number of 'old' systems outnumbers the 'new' systems many times over. In the same way that it's impossible to exist without being affected by an IT system, it's impossible to avoid legacy ones. Even if you have the world's largest group of developers and an infinite budget in a brand-new organization, you'll still have to integrate with legacy systems and eventually your own green-field projects will become legacy.

If you have a position of responsibility within an organization, you will have to deal with legacy systems.

However, the IT industry is obsessed with new technologies and new projects. University courses, books, magazines, and conferences focus on what is new and assume you always start with a clean slate. This isn't what occurs in the real world, and I hope what follows fills some of this gap. This book is *not* intended to present a formal methodology but is aimed at all the Penelope's out there who need a guidebook to help them with their first legacy system.

Robert Annett (**robert.annett@codingthearchitecture.com**)

About the Authors

Robert Annett has been a developer since 1995 and has worked in industries from energy management to investment banking. Much of his work has involved upgrading and migrating legacy systems, with the occasional green-field project (involving integration to legacy systems, of course). He has worked in the IT industry long enough to realize that all successful systems become legacy eventually.

Learning Objectives

- Perform static and dynamic analyses of legacy systems

- Implement various best practices to secure your legacy systems

- Use techniques such as data cleansing and process cleansing to stabilize your system

- Apply structural changes to your legacy system to make it highly available

- Identify and resolve common issues with legacy systems

- Perform various tests to secure your legacy systems

Audience

Working with Legacy systems is ideal for IT professionals who want to understand the workings and maintenance of legacy systems. Prior knowledge of working with legacy systems is not needed to read this book.

Approach

Working with Legacy systems takes the learning-by-doing approach to explain concepts to you. It uses practical and hands-on practice sessions to understand the workings and maintenance of legacy systems. Prior knowledge of working with legacy systems is not needed for this book.

Acknowledgements and Thanks to:

- Simon Brown for encouraging me to write this in the first place
- Richard Jaques for reviewing an early copy of the book, making suggestions, and pointing out a few embarrassing typos
- Robert Smallshire for reviewing an early copy of the book and giving direction on the target audience
- Graham Lee for reviewing an early copy of the book and keeping me amused via Twitter
- David Hayes for the review and comments on structure

1

Definition, Issues, and Strategy

What is Legacy?

For most people, in most situations, being passed a **legacy** is a good thing. Perhaps their long-lost uncle has left them a stately home filled with antiques. Or a legacy might be a valuable art collection given to a museum or the work of a famous author. However, in Information Technology, it is a dirty word filled with innuendo and often used as an insult. This is a strange situation and we should define exactly what we mean by legacy before addressing how to deal with it – and even if we *should* deal with it!

When we refer to a system as being legacy what we're really saying is that the system is built in a way that differs from how we'd choose to do so today. Legacy systems may have been written well, using the best technologies and tools available at the time, but those technologies are now out-of-date. A system written in 2001 using Java 1.2 and Oracle 8i may have made perfect sense but if you wrote it now, you'd at least use the latest versions available or even something different entirely (**Scala**, **MongoDB**, and so on didn't exist then).

An IT-Centric View of the World

Like most developers, at the center of my working life is the software development process. This might initially involve a business analysis and some support afterward, but this takes up a relatively small percentage of my time.

A **software-development-centric** view of the world is as follows (this is the **Software Development Life Cycle** (**SDLC**) diagram and your own process may differ slightly). There are two parts that interact with the outside world – **requirements** and **deployment** (which may be lightweight and frequent) and external interactions are wrapped in those phases.

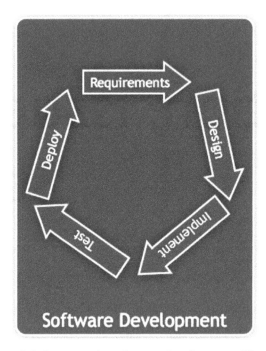

Figure 1.1: An example software development life cycle

For some businesses (for example, public-facing websites), software is the business and there is no significant period before or after software development – it occurs constantly. These businesses are very interesting to us developers for this very reason (and therefore, these projects are often discussed at conferences); these are the exception. For most organizations, their individual IT systems perform a function that constitutes only a small part of what the organization does.

Systems Development in Context

Therefore, software users view the world differently. The software development phase is a very small part of their business processes, life cycle, and lifespan. They view the world a little more like this:

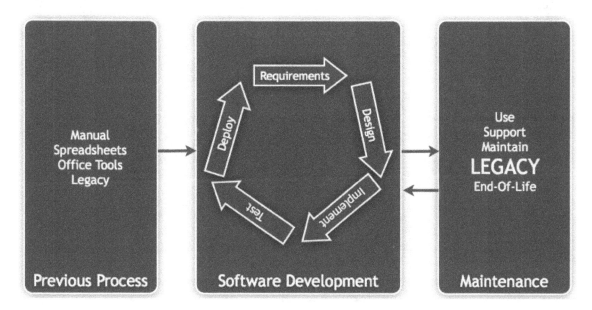

Figure 1.2: The life cycle in context

Most of the processes the software users execute will originally have been done manually (I include generic software tools such as spreadsheets) and may have been done this way for 6 months to 100 years. (If you think I'm exaggerating with "100 years" then you should speak to someone who's worked for an old insurance company.)

At the end it will be decided that it is cost-effective to develop bespoke software (or spend a lot of effort configuring BPM/CRM software, and so on). This process may be iterative and deliver value quickly but will be mostly complete after a relatively short period of time (compared to the organization's lifespan).

The software is now fully live, and a software team will consider it to now be in its maintenance phase. From the organization's point of view, this is where the real value is extracted from the system. Very little money is being spent on it, but it is being used and is either helping to generate revenue or increasing the efficiency of a process.

As time goes on, there will be a decreasing number of changes made to it. Eventually, it will be considered legacy. Ultimately, the system will reach end of life and be replaced when it is no longer fit for its purpose.

Systems Development Scaled with Time

If we were going to scale the diagram to indicate time, then it might look like this:

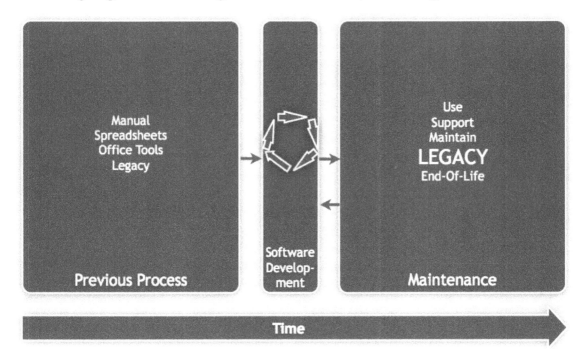

Figure 1.3: The life cycle in context and scaled with time

These are typical life cycle times that I've experienced and yours may differ – this will depend largely on the industry you work within.

Therefore, legacy should be viewed as a phase in the life cycle of the process. Note that I have a small arrow going back to Software Development from the maintenance/legacy phase, as it may have upgrades or additions many years after being deployed.

> ### Line-of-Business Systems versus Others
>
> It was pointed out to me (thanks to Robert Smallshire) that the life cycle I've described best fits line-of-business software development and other types of systems that may vary greatly (for example, the control systems for embedded devices). This is true and reflects my personal experience but the statements on differing perspectives still hold.

I wrote a blog post covering much of this and someone commented that they refer to legacy systems as mature, as this has a more positive connotation. I think it also accurately reflects legacy as being a stage of a life cycle rather than any indication of quality.

A system maintained by professional programmers would be periodically upgraded and migrated to new technology when required, so it may never be mature/legacy. However, this is rarely a decision for programmers – it's a management decision. To be fair to management, does it really make sense to spend large amounts of money continually upgrading a system if it works? If the programmers and systems team did a good job, then no changes may be needed for many years. This indicates a high return on investment and a good quality system – a legacy!

Of course, when you do have to make changes or additions, you have a very different set of issues compared to modifying a system that is being continually developed.

Examples of Real Legacy Systems

Giving an example in a publication is dangerous as, in five years, it might be hopelessly out of date. However, considering the subject matter, I shouldn't make this argument. Look at the following:

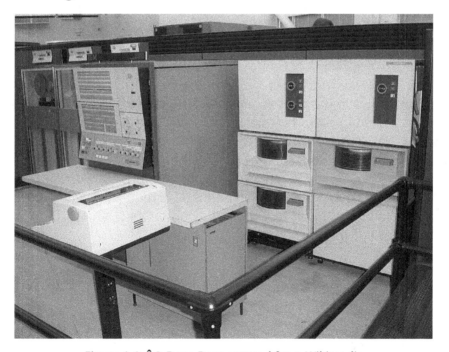

Figure 1.4: Â© Dave Ross sourced from Wikimedia

Is this a legacy system? It's a museum piece. This is an IBM360 and was once the most popular computer on the planet. However, there are no commercially working examples left and they can only be found in museums.

This shouldn't surprise us, given the example timeline and arguments from earlier. Systems will be end-of-life or modernized eventually and it's very rare for a working system to be this old. So, what are the types of legacy systems you're most likely to come across? This will obviously change with time (and depends on the organizational areas you work within), but they are most likely to be:

- Between five and fifteen years old
- Without maintenance for two years
- Using out-of-date APIs or technologies

If they don't meet these criteria, then you are unlikely to consider them to be legacy. Some examples are:

- Java 1.2/1.3/1.4 running under Solaris 8 on an Ultra Sparc III server using Oracle 8i
- C# 1.0, Windows server 2000 using SQL Server 2000
- Objective C on NeXT server
- J++/J#, VB6, FoxPro, Brokat Twister
- A combination of spreadsheets, macros, and shell scripts

I'll cover common issues in the next section, but the preceding examples have different problems even though they are technologies from between 10 and 15 years ago. The first two (Java/Oracle and the **.net** stack) are very out of date but are all supported technologies. These should be simple to upgrade and support but are likely to be challenging to modify. The next two involve hardware and software that are discontinued. This will be much more challenging. Lastly, we have the very common situation of a system being bolted together from 'office tools.' In this book, I'll be covering these kinds of legacy systems.

> **Note**
>
> I worked for an investment bank that had made a large investment in systems written in Objective C and running on NeXT servers. These systems worked well, but were complex and hard to replace. They ended up being run for so long that they had to start buying hardware spares off eBay.

Common Issues

In this section, I'll be identifying common issues with the types of systems discussed in the previous section. Depending on previous experience, these might appear obvious (you've seen it before) or surprising (you'll see it eventually). I would suggest that when you take on a new legacy project, you go through these and see what applies. Being aware of and expecting these issues will help you deal with them. I have included an *Issues Checklist* in the appendix to help with this.

No Documentation

When a system has been running for a long time, it is not unusual for supporting artefacts to be lost. The users of the system will notice if it stops working and will complain, but will anyone notice if items, such as documentation, are lost? This can happen for several reasons:

- Support systems may be retired, and information removed. This can happen when there is no obvious link from the documentation to the system it concerns.

- Documentation still exists but no one knows where it is or if it has been moved and the links/locations are not valid. This is a common problem when migrating document management systems.

- Information is in an uncommon or unknown format. It's amazing how many files end in **.doc**.

- Information was only ever stored locally (and the machines have been wiped) or not stored at all.

The agile manifesto says we should favor *working software over comprehensive documentation*, but it does *not* say we should not document anything. There are various types of metadata about our system that can make supporting it very difficult if missing. For example:

- **Usernames and passwords**: The direct users of the system and its administrators may know their access details but what about the system's sub-components? What about administrator access for the database or directory server? If these are lost, then performing maintenance actions can be incredibly difficult.

- **Release instructions**: Maybe you have the source code, but do you know how to build and release the software? Data is more likely to be kept than a running, unused service so your build and deploy server probably won't be present after 10 years. Was your build configuration in the source control system?

- **Last release branch details**: Which one of the multiple code branches or labels was released?

- **Communication protocols**: This is one of my personal bugbears. Many systems have been designed over the last 10 years to communicate with each other via XML messaging. Whereas developers used to document binary message formats, they've often not done so with XML as it's deemed to be 'human-readable' just because it's text-based. XML data blocks are rarely obvious, and they have so many optional elements that they are very difficult to reverse engineer. This is especially true if you only control one side of the communications, for example, receiving a message from an external source.

- **Licenses and other legal agreements**: I'm going to talk a little more about licenses later, but can you track down the legal information about how you can use the third-party software and hardware elements in your system? They may be more restrictive than you think.

- **Users, including external systems**: Do you know who your users are? What external systems rely on the one you control?

It is always worth tracking down and cataloguing all relevant documentation as a first step to maintaining a legacy system.

Lost Knowledge

There is an overlap between lost knowledge and a lack of documentation. You could argue that no knowledge would be lost if everything was documented but this is unrealistic and incredibly time-consuming. It's likely the reasoning behind design decisions has been lost. You have the artefact (the system) but not the how and the why as to its creation.

> **Note**
>
> Many years ago, I worked on a system that, once a day, collected some customer information and emailed it to some users. As part of the process, a file was created (called `client-credit.tmp` in the `/tmp` folder on a Unix machine). The next day, this file was overwritten. When we began upgrading it, we moved the process into memory, so no *temporary files* were created. I received a very angry phone call a few days after go-live from someone whose system was broken. It turned out that another system (in a totally different part of the organization) was FTP'ing onto the server, copying `client-credit.tmp` from `/tmp` and using it. At some point, there must have been a discussion between the two departments about giving access to this information and how to do it. However, once this hack was implemented (maybe this was not supposed to be the permanent solution) it was forgotten and those responsible moved to different jobs. It was interesting that no one from either department knew this was happening.

These types of lost knowledge can create some nasty traps for you to fall into. Some of the techniques and processes discussed in later sections should help you identify and avoid them.

Hidden Knowledge

Also referred to as **Special Voodoo**, these are useful but obscure pieces of information that may either work around a bug or initiate a non-obvious function. Although not *lost*, it is also not documented and deliberately hidden by an individual or small group of operators. Why would anyone do this? The most common answer is *job security*. If you're the only person who knows how to bring up the system when it freezes or create a certain report, then it makes you much harder to replace.

This is very, very common in legacy systems and can mean that the very people you need to help you support and improve the system might be actively working against you. I'm going to be covering this a little more in a later section on *Politics*.

> **Note**
>
> I once worked for an **ISV (Independent Software Vendor)** that had written a complicated piece of analysis software. This had started as a simple tool and had grown organically over 10 years and was now a system used by large organizations. There was an operator for this software at one site who would 'work from home' 1 day a month in order to complete a monthly report – which he presented when he returned the next day. His colleagues (and boss) thought that he was working all day on the report, but he spent it playing golf. The report only took 10 minutes to generate by exporting the data from an obscure part of the system and then loading it into a spreadsheet template.

Unused Functionality

In a large system that performs complex functions, there is a good chance that there are huge chunks of functionality that are not used. This tends to be either due to an over-zealous development team that added features that have never been used (over-engineering) or features that are now unneeded. Often, parts of the system are replicated elsewhere in newer systems and we can think of our legacy system as being partly replaced.

Why do I describe this as a problem? Can't we just ignore these? If you don't know whether part of a system needs to be supported or maintained, you might waste a long time trying to migrate something that is not used. However, if you decide that part of the system isn't required, and then do not support or maintain it, you might later discover that it was just used very infrequently.

The owners of these rarely used features might be difficult to track down, making it hard to get a sign-off for changes.

Some unused features might still be required even if you never use them or ever intend to use them. This is actually very common in software used in a regulated or safety critical environment. For example, having the ability to delete records for security reasons (privacy regulation) or features used in disaster scenarios. You shouldn't turn off the control systems dealing with the meltdown of a nuclear power station just because it hasn't been used in 20 years. Unfortunately, there are examples of safety systems being removed in just this way – it's important you know this.

Even if your system doesn't have a potential catastrophe, you should consider whether the **Business Continuity Plans** and **High Availability** systems still work.

Figure 1.5: Old tractor image

Is this tractor used? It's old but not abandoned. It might start but is it used to do any work? Perhaps it's the only tractor in the farm that has a certain attachment to drive an essential, but rarely used, piece of equipment? Importantly, if you're the mechanic on the farm do you spend a huge amount of time keeping this working? Can you even find anyone to ask about this?

No Coherent Design/Inconsistent Implementation

This can be due to an active decision to avoid top-down design but is more often due to a system growing in an organic way.

Many successful, large systems don't start as large systems. They start as a useful tool doing a difficult but specific task. If they do the job well, then they'll have features and options added. Over a long period of time it can morph into a large and complex system with no coherent design. Common indicators of this are:

- Business logic in the incorrect place, such as in a client rather than a service. Perhaps this started as a standalone application that morphed into a multiuser application.

- Multiple tools for doing the same job – for example, different XML parsers being used in different parts of the system or several ways of sending an email report. This is often due to a new developer bolting on a feature using the tools they know without looking at the rest of the system.

- Inappropriate use of frameworks. For example, using threads against advice in an application server or trying to get a request-response style service to schedule tasks. This is normally due to adding a feature that was never originally envisioned and trying to do it in the current framework.

- Many layers and mapping objects. New features may not fit into the current 'design' and require a lot of supporting code to work with what is there.

- Sudden changes in code style – from pattern use to formatting. Where different developers add features at different points.

Some of these may be bad (time for a refactor) but for others, it may just be inconsistent. Why does this matter?

Lack of coherence in design and consistency in implementation makes a system much harder to modify. It's similar to driving in a different country – the road systems may make perfect sense, but they take a while for a foreigner to get used to. And if you move country every week, you'll never be a good driver again. Similarly, it does take a developer a while to get used to patterns and tools and they will be much less productive in an incoherent system.

It's much harder to predict the non-functional behavior in such a system. If each part of the system works differently, then performance tuning becomes hugely complex.

Fragility (versus Stability)

The first thing to note is that I've used the term **fragility** rather than **instability**. The difference is subtle, but important. An unstable system will stop working in a frequent and unpredictable way (if it was predicable then you'd have specific bugs). Whereas a fragile system will work as required but will crash horribly when a change is made.

Figure 1.6: It's stable but fragile. Don't open a window

Fragility can be worse than instability from a maintainer's perspective. If the users consider a system to be fragile, it can be difficult to get permission to modify it. Sometimes this fragility may be perceived rather than actual, and this is common when the system is old and no longer understood – the fear is that making any changes will stop it from working and no one will be able to repair it.

In contrast, sometimes an unstable system can work in your favor, as you can make mistakes, and no one will notice.

Later in the book, I'll be covering some of the tools and techniques for making changes to fragile systems.

Tight Coupling

In a **tightly coupled** system, it can be very difficult to pull apart the system's different components. This makes it hard to upgrade in a piecemeal fashion, leaving you with a stressful 'big bang' approach to upgrading and maintenance. It's also very difficult to performance-tune problematic functions.

Loose Coupling

What? Surely **loose coupling** is a good thing? However, what about the anecdote I told earlier about the temporary file that was FTP'd by another service? Although this is a terrible design, isn't it an example of loose coupling?

With a tightly coupled system, it's usually obvious what the dependencies are, and nothing runs without them being in a specific state. With a loosely coupled system (particularly one where services depend on each other and communicate asynchronously via a bus) this may not be so obvious. This is particularly problematic when there is a lot of unused and rarely used functionality (see earlier issues), as you don't know how or when to trigger potentially important events.

Zombie Technologies

The reason these are called **Zombie** rather than **Dead** is that when I investigated some example technologies, I found out that almost no technology dies. There is always some hobbyist somewhere insisting that it's still alive, therefore, I'd count the following as zombie (dead in reality) technologies:

- Technologies where the company that created and supported them no longer exists

- Technologies no longer officially supported by their creators (including open source projects that have not been touched in years)

- Technologies not compatible with the latest versions

- Technologies where it is not possible or practical to find engineers that have any knowledge of them

- Technologies where important parts have been lost (such as the source code)

I'm sure I'll get complaints about some of the preceding definitions, but these are all situations that make a system much harder to upgrade or maintain. Trying to solve a bug or modify the deployment for these can be difficult and modifying functionality might be impossible.

We should also remember that fashions and the way technologies are used can change a lot over time. Well-written Java code that used best practices from 2001 will look completely different from code a Java developer would write today.

Licensing

All the components in a system, both software and hardware, will be covered by a **license**. This is something very often forgotten by technologists when inheriting a legacy system. Licenses can say almost anything and might put very restrictive conditions on what you do with components. This applies to everything through the entire stack and potentially even some of the data in the system if it was supplied by a third party with restrictions on its use.

Here are some questions you should ask about a system and its licenses:

1. Can you make changes?

2. Does it stop you using virtualization?

3. Are you not allowed to run the software on certain hardware, or does the hardware only run certain types of software?

4. Is **reverse engineering** prohibited and will this affect your ability to instrument it?

5. Can you modify any source you might have?

6. Can you even find the licenses?

7. Is any of the data owned by a third party?

The cost implications of not understanding licensing can be high. In the January 2013 issue of *Computing Magazine*, some of the implications of licensing and virtualization were discussed and the following quote was given:

> ### Hidden Licensing Costs
>
> "If a company has SQLServer on VMware, licensed on a per core basis that can be dynamically scheduled across hosts – all machines need licenses." – Sean Robinson, License Dashboard quoted in Computing.

As well as problems regarding restricted use, licenses are also a **sunk cost**. If an organization has paid a large amount of money for a product, then it can be politically very difficult to change, even if the alternative has a lower overall cost.

> **Note**
>
> **Sunk Costs** are retrospective (past) costs that have already been incurred and cannot be recovered. It is an interesting area of behavioral economic research. In theory, a sunk cost should *not* affect future decisions – it is already spent and can't be recovered. In practice, people find it very difficult to 'write-off' previous expenditure, even if this makes economic sense.

Regulation

Regulation is very similar to licensing in that it also concerns external rules imposed on a system that are outside the normal functional and non-functional requirements. Although licenses stay the same when you want to change the system, the difference with regulation is that it tends to change and force functional changes on you.

Shortly before writing this book, there have been changes to many websites in the UK. A huge number have started informing users that they store cookies and requesting permission from users to allow it. This is due to the **Privacy and Electronic Communication Regulations (PECR)** Act 2011 which is the UK's implementation of the European Union's ePrivacy laws. It's often referred to as **The Cookie Law**.

Of course, not all websites have implemented this and the ones that have are the high-volume websites with permanent development teams. Many smaller organizations with legacy websites that aren't actively developed, have not started asking these questions and are arguably not compliant. I'm not a lawyer though, so seek legal advice if you think you have a problem!

It's interesting that so many organizations have taken this regulation so seriously and it's probably because the use of cookies is so easy for an individual outside the organization to test. We can imagine someone writing a script to find UK-based websites that don't comply and then trying to sue them for infringing their privacy rights.

There are many other regulations that affect IT systems, from industry-specific ones (finance, energy, and telecoms are heavily regulated industries), to wide-ranging, cross-industry ones, such as privacy and data protection. Regulations regarding IT systems are increasing, and systems normally must comply even if they were written a long time before the regulation came into existence.

Please remember that failure to comply with regulations may be a criminal offence. If you breach a license agreement, you might find yourself being sued for a large amount of unpaid fees, but if you fail to comply with regulations (for example, money laundering or safety reporting), the consequences could be much worse.

Politics

Technology workers are often very bad at spotting political factors that affect their projects. Here are some questions you should ask yourself about your project (whether legacy, green-field, or somewhere in-between).

- Who owns the project?

- Who uses the project?

- Who pays for the project?

- Who gets fired if it fails?

- Whose job is at risk... if successful?

What we should be looking out for are conflicts of interest – so let's look at the last two I've listed.

If a project fails, then the sponsors of the project and the implementers of the project will take the blame and their jobs may be at risk. This is to be expected and is one of the reasons we're motivated to do a good job and why we get harassed so much by project sponsors.

However, we rarely consider who suffers when a project is a success. It's much easier to get funding for a migration or upgrade if there is a defined business benefit; even easier still if there are increased revenues or costs savings (profit!). When an IT system saves money, it's often because jobs can be **rationalized**, that is, people get fired.

> **Note**
>
> Once, I was taken to see a client by a salesman for the software company we worked for. We walked through a room of data entry clerks (most of whom were women) and, being a stereotypical salesman, he stopped to flirt several times (he already had a couple of divorces and was working on his third). Once we left the room, he said "It's a shame isn't it?". I was a little confused and asked what he meant. His reply was "Well, once we've installed the data-importing software you've written, they're all out of a job". I hadn't been out of university for long and was naive to the real-world effects (and potential politics) of what we were doing.

Even if jobs aren't made redundant by a system improvement, the user might suffer in other ways. If tasks become standardized/commoditized, then it's much easier to replace a specific user with someone else who is cheaper (outsource the processes) or has a better attitude. Remember the examples of hidden knowledge I gave earlier? An improved system might remove this advantage for power users.

We must remember that many people simply don't like change. Technology workers are very unusual, as we like change; it's often what drew us into the job in the first place. We relish the way we work changing and making ourselves redundant probably holds no fear, as we move jobs every 18 months anyway. The users of a legacy system may have been doing the same job in the same way for 10 years and the thought of learning something new (and possibly being bad at it) can scare them.

You should also question the costs and benefits. There is a good chance that the cost is incurred by someone different to the beneficiary of the project. This is often the case in larger organizations where the IT department is separate from the business units using the software. Different solutions to issues may incur costs to different groups. For example, a re-write of a piece of software may come out of the business unit's budget, whereas an 'upgrade' might be viewed as the IT department's cost. Virtualizing an old system will probably be a cost to the IT department but the cost increase or savings on business software might fall to the end users. This will cause conflicts of interest and the decisions taken may be based on the political power of the department leaders rather than being the best decision for the organization.

Politics can be very destructive, and I've seen this as the cause of failure of many projects. From personal experience, I'd say that it's more often the cause of failure than bad technical decisions or poor implementation.

Organization Constraints and Change are Reflected in the System

Beyond the deliberate effects of politics, there are more subtle marks that an organization can leave on the systems that are developed for it. This is described as **Conway's Law** (http://en.wikipedia.org/wiki/Conway's_law).

Not only are these organizational constraints reflected in the system when developed but these constraints will change over time. As people leave, teams merge or split and the barriers that existed may disappear or new ones may be created. However, the artefacts in the system will remain and build up over time. This can become obvious in long-lived legacy systems.

This affects software and hardware. Consider a system that two different departments in an organization are using. They may want separate databases (that only they have access to) and may want their processes/reports to run on servers they own (perhaps they had to provide some budget for the system's creation, and they want exclusive use of 'their' server). These departments may change dramatically over time, but you'll still have two databases and two servers.

Understanding the organizational structures of software and business teams can help to explain which quirks of the system are due to this and which are due to some large, complex subtlety that you're yet to understand (hat tip to Richard Jaques, whom I stole this sentence from).

External Processes Have Evolved to Fit around the System

An IT system is part of a larger process and workflow. It will have been developed to fit into the process to make it more efficient and effective. However, it doesn't just fit into the outside world; it will also affect it. The processes surrounding the system will change and evolve to work with it and its quirks. With a legacy system, these processes will have become embedded in the organization and can become very difficult to change (remember that people don't like change).

This can lead to the peculiar situation where you must get a newer, better system to work with old and inefficient processes that were originally imposed by the proceeding system. When gathering requirements for a new system, you should remember this and make sure you are finding out what is *needed* rather than what is *currently done*.

External Systems Have Evolved to Fit around the System

The external systems that interact with the legacy system will have also evolved to match its interfaces. This is especially true if the legacy system is quite old and the external systems were created afterwards.

This can lead to a situation where a replacement system must match old and unwanted interfaces in order to interact with dependent systems that only have those interfaces to connect to the legacy system that is now being replaced. Ancient interfaces can therefore survive long after the system they were originally designed for has been removed.

Decaying Data

I've already described how systems can grow organically, but the data within them can also decay (The metaphor isn't perfect but I'm trying to make this interesting). This is a decrease in the overall quality of the data within the system caused by small errors gradually introduced by the users and external feeder systems. Examples include:

1. Data entry (typing) errors

2. Copy-and-paste reproduction errors such as missing the last character or extra whitespace

3. Old data, that is, the details are no longer true, such as an address changing

4. Feeder systems changing formats

5. Data corruption

6. Undeleted test data

When a system is new, there are often data quality checking tasks to keep the data accurate, but with time these tend to be dropped or forgotten (especially tasks to prune data no longer required). This is another example of something from behavioral economics, called *The Tragedy of the Commons*.

> **The Tragedy of the Commons**
>
> When many people have joint use of a common resource, they all benefit from using it. The more an individual uses the resource, then the more they personally benefit. If they mistreat it or overuse it, then it may degrade but the degradation is spread amongst all the users. If an individual spends time improving the resource, then any increased benefit will also be spread amongst the whole group. From a purely logical point of view, it makes sense for any individual to use it but not spend any time improving the resource. They are acting independently and rationally (according to self-interest) but the resource will degrade over time.

These errors in data can (and sometimes without being noticed) accumulate until the entire system becomes unusable.

Now What?

Did you recognize any of those issues from previous projects you've worked on? I've not listed the system being 'bad' as a problem, as it might not be perfect (all real systems have issues and bugs), but if it's a legacy system then it must have value, or it would just be turned off.

When working on a legacy system, it's worth recording issues and potential issues, such as those listed previously. Consider starting a 'Problems and Issues' document for your legacy system and recording real and potential problems. Please see *Appendix 2 - Legacy Project Questions* as a potential starting point for this.

Reasons to be Cheerful

I spent the last section identifying problems, but I want to be positive about a situation we often find ourselves in – and I think that some of the issues with legacy systems are due to neglect caused by negativity. So, here are a few of the reasons why working on a legacy system should be viewed as a positive experience.

A Legacy System is a Successful System

I've already mentioned this but it's worth repeating. If the system had no value, then it would be turned off. It is worth finding out what the value is, even if the only reason you do so is so that you can make sure a replacement system is at least as good. Many of the positive points might only be obvious after a good investigation. Users will be very forthcoming with complaints but the parts they don't talk about can be excellent.

You Have Real Users You Can Talk To

One of the greatest obstacles to developing a **Greenfield** project is knowing who the end users will be and what they require. With a legacy system, you have end users you can speak to. They can tell you what they like and dislike about the current system and tell you what they would like to be included in any upgrade. This is a massive advantage but is also one that many developers don't take advantage of.

You Can Learn a Lot about the Business

The system might define a lot of the business and organizational processes. Many of the business decisions will be automated in the system and much of the external interaction. Some of the more complex decision processes might have been carefully defined several years before and now no one is left who knows them. The only place you can learn is by examining the IT system! I have found this to be particularly true in the finance industry. You can learn enough from an existing system to ask intelligent, relevant, and specific questions of business users (for example, the system is doing X, should it also be doing Y) rather than vague, open questions they may find difficult to answer.

You Can Have a Large Impact Quickly

If a system hasn't been modified for a while, but the way it is being used has evolved, then small changes can be made that help the users a lot. I'll talk about this more in later sections but simple performance tweaks (such as adding an index so a report runs in good time) or automating a simple, manual process can dramatically improve the user experience. You can look like a hero for little effort!

It's Important, But Not Trendy, So You'll Get Paid Well!

Everyone wants to work on the cool new system with a cutting-edge graphical interface on an expensive mobile device. The law of supply and demand means that the cost (pay) goes down for this popular work. A high value system that isn't trendy means you can justify being highly paid. Maybe this is shallow, but if you have a mortgage and a couple of kids, then you'll understand.

Strategies

Once you have inherited a legacy system, what do you do with it and how do you approach any issues? Most books on software and system design assume it is a Greenfield project where you can build how you please. Legacy systems have constraints on what you can do and how you can do it.

This section suggests some strategies for managing these systems.

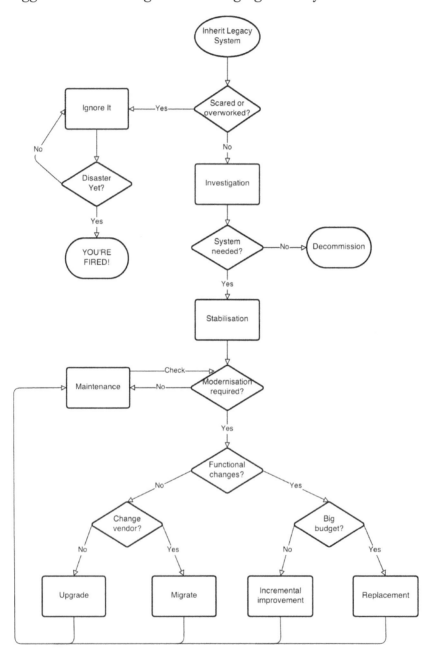

Figure 1.7: Strategy Decision Flowchart

Ignore It

What is it?

I'm not being facetious, but I've had to include this as a possible strategy as it's the most common! If you choose to ignore your legacy system, then you should be aware that you've still made an active decision – which is to make no changes and not improve knowledge about any issues. You're not fooling anyone if you say it's on a long-term plan either.

Ignoring a legacy system is very tempting because of the issues and problems we discussed in previous sections. A lack of documentation and lost knowledge will make any project or investigation hard to even start and the politics surrounding it might make you unpopular.

Advantages

The only advantages to ignoring a legacy system is that you'll avoid any political battles in the short term and can get on with other work.

Disadvantages

The risks to ignoring the system are huge. We've already seen some of them, such as regulatory changes and zombie technologies. We also must remember that hardware, degrades and old processes might not work as expected any longer. We could argue that the biggest risk is the lost opportunity of not improving processes that could make the organization more efficient.

Investigation

What is it?

Performing an investigation of your legacy system's structural components, runtime attributes and usage – that is, what makes up the system (hardware, software and configuration); how the system runs (inputs, outputs, and so on); and the use cases it is involved with.

Although this is a strategy (as you may discover that much of your system requires no further treatment or can be decommissioned), it should also be viewed as a stage to be performed before the other strategies.

> **Note**
>
> I worked on a project to rationalize a system that was made up of a set of components linked together by some manual processes. One process involved a user who received a file via email, which they then uploaded into an internal web application. This application performed some actions on the file and then dumped it to a central location. The users then loaded it into another system. On investigation, it turned out that all the web application was doing was changing the line endings of the file from Unix to Windows. However, the system this was loaded into would cope with either style of line endings. We had no idea why the intermediate step was first introduced (presumably, it was required at one point and then became part of the 'ritual') but it meant we could drop that part of the process and turn a sub-system off. There were good cost savings as the department was being charged a reasonable sum by centralized IT to host the web application.

You might perform a detailed investigation before deciding upon a strategy or this may have already been decided for you (we'll discuss migrations later, which are often cost- or politically-driven). Even if your strategy is predetermined, the investigation phase is important but is often passed over by keen workers desperate to show progress – this is a mistake.

Advantages

This is a good way to start working with a legacy system and shouldn't incur much political opposition as no decisions on actions are taken yet. There is also a good chance that, after an investigation, you'll change your preferred strategy – often because you've discovered value in the system you weren't aware of.

Disadvantages

You can become stuck in 'analysis paralysis' if the system is complex and opaque, so you should time-box your initial investigation. It can also be difficult to spend the required time performing the investigation properly if other commitments with higher priority repeatedly bump the work (investigations are usually ranked as a low-priority).

Advice on Implementation

An investigation should involve all the planning steps that are detailed in Part II and you should aim to generate a set of artefacts that can be used to facilitate a discussion of the system and future maintenance and upgrades.

When performing the investigation, you should also ask the following:

- What is being done?
- How is it being done?
- Why is it being done?
- Is it necessary?
- Is there an alternative?

These are leading questions (looking for possible de-commissioning) but you are trying to get a deeper understanding of the system rather than just capturing it. It will also help you understand any value that it contains.

Maintenance

What is it?

All IT systems need maintenance, unless you intend to decommission any unused systems immediately. By maintenance I'm referring to keeping the system in its current state and correcting problems, rather than introducing functional or non-functional changes and improvements. If you intend to introduce improvements, then this would be considered upgrading, migrating, or possibly incremental improvements.

Rather than trying to improve the system, you just want to stop the system from degrading. Real systems degrade with time so monitoring and maintenance procedures are necessary to stop this from occurring. Examples of maintenance might include:

- Operating system patches
- Essential third-party security patches
- Replacing failing hardware such as hard disks with sector errors
- Removing unused users
- Rebuilding indexes to maintain performance as data is modified
- Deleting temporary files to stop a file system filling

You are simply trying to 'sweat the assets' and avoid any catastrophic issues. Hopefully, when the system was built, a set of maintenance actions was documented but this may not be the case; they might have been lost or evolved over time. You may require a large investigation stage, and if maintenance has been neglected, a substantial stabilization task (covered later) before resuming with regular maintenance.

Advantages

There are many benefits to maintaining a system rather than upgrading it. For example, you may work in an environment where systems require large amounts of security investigation or sign-off for functional provability. NASA is rumored to have kept systems from the 1970s running the space shuttle program because the overhead of proving reliability was too high to ever perform major upgrades or outright replacement.

Keeping a system's functionally static may also help to avoid many of the political issues that we listed in the previous section.

Disadvantages

When a system is receiving only maintenance, all the subcomponents will gradually age, as they are not being upgraded. As hardware components approach their expected lifespan, they may fail. Eventually, these hardware components will need to be replaced but replacements can be difficult to find for old systems. Eventually, an upgrade will become unavoidable, but this is best done at leisure rather than in a panic after a failure.

Software ages in a different way and it may no longer be supported, compatible, or known by new staff members.

Many projects in the maintenance phase end up morphing into other projects. It's tempting to gradually add features into a tactical system while a strategic system is being developed, which often means the strategic system never catches up.

Advice on Implementation

Scheduling investigation, stabilization, and maintenance as individual tasks will help with planning and estimation. Telling the business owner that these are separate will help with budgets and timescales (more politics). (By business owner, I'm referring to the individual or group that is the sponsor for the project. They are usually the budget holder and therefore have a lot of influence over the project. This includes organizations in the public sector and not just commercial businesses.)

You should resist allowing a business owner to 'just add a little feature' if you've agreed to maintenance only, as any scope creep will cause issues. Adding features involves another strategy, which is approached differently. I would advise getting stakeholders to agree to this strategy up-front so that, if new features are requested, you can justify extra resources.

Upgrade

What is it?

This strategy involves upgrading each component of a system with the latest, supported version. This may include:

- Operating system (for example, Windows 2008->20XX)

- Versions of Java/CLR (for example, JDK1.3->JDK1.8)

- Database (for example, Oracle X -> Oracle X+1)

- Processor architecture (for example, newer/faster/more processors)

- Communications infrastructure (for example, 100MB NIC -> 1GB NIC)

- Messaging infrastructure

- Third-party products such as CRM/BPM systems

The aim is for little functional or non-functional change (although, for some third-party business-process components, this may be unavoidable).

The hardware and software development life cycle make it difficult for vendors to support old versions. Therefore, if your organization is using very old third-party software or hardware, then the original manufacturer may no longer provide support and force you into upgrading. Even if your organization wrote the main applications you are using, there will still be frameworks and infrastructure affected by this.

A strategy that involves new functionality may involve an upgrade step first.

Advantages

Upgrading current components allows users to continue using a product they are familiar with. The product should be easier to maintain as it will be better supported. Components that are upgraded to their latest versions are more likely to comply with regulations and standards, and they are often viewed as 'best practice'.

There are important cost implications for both support (out-of-date products often incur higher support costs) and in comparison, to buying new products (upgrading software should be cheaper than entirely new software and upgrading hardware components should be cheaper than an entire new system).

Disadvantages

Upgrades are rarely as simple or straightforward as they should be – here are a few issues you might face:

- Dependencies between components

  Performing an upgrade on a single component is often impossible. You may want to upgrade one specific component (perhaps due to known issues, such as security) but there are often dependencies between components that force you to upgrade others. These dependencies are both vertical and horizontal.

  > **Note**
  >
  > Recently, I had to upgrade a piece of third-party software due to the version entering end of life. This software came with a much later version of a database driver, which didn't work well with the old version of the database we were using, so we needed to upgrade it (horizontal dependency). The latest version of the database required a lot more memory and would only work correctly in a 64-bit configuration. This required us to upgrade the operating system on the database machine (vertical dependency) and the combination of the new database and OS meant that we needed new hardware, which we virtualized to manage better.

  An upgrade to a single piece of software necessitated upgrades across multiple components right down to the hardware and introduced extra layers.

- Issues with data

  I mentioned decaying data in the *Common Issues* section, and bad data creates specific problems for upgrades.

It is common for newer versions of software to assume that the data is of higher quality than was required or guaranteed by earlier versions. The operators of a system will use it in a way not originally envisioned by its creators and will insert data that is not specifically excluded. For example, adding comments to a telephone field because the software didn't enforce numeric values only. When the software vendors tested the upgrade process, they probably used sensible and expected values rather than the values found in the real world.

This unexpected data may cause repeated upgrade failures or, even worse, might create unexpected behavior in the system.

- Several versions at once

If the legacy system has been in maintenance only (or ignore) mode for a while, then components of the system may be several versions behind in the upgrade cycle. It may not be possible to directly upgrade software components from your production version to the latest and you may have to move to an intermediate version (assuming you can even get hold of one). This can introduce a large amount of testing. See later sections about using virtualization to help with upgrading, as you'll want to take snapshots at these intermediate points so you can roll back to positions other than your starting point.

Sometimes, if the version you are upgrading is old enough, then you are really performing a migration and you might want to skip the intermediate versions. In this case, you would perform a fresh install and then migrate and import the data.

When you're asked to upgrade a component, you should be very careful to consider the true time required. Upgrading an item in a legacy system may be time-consuming and involve work in unforeseen areas. It can be very different to upgrading part of a non-legacy system. Please also remember that, after an upgrade, you'll still need to maintain the system – don't upgrade and forget.

Advice on Implementation

I've already listed some specific issues with upgrading and some advice for dealing with them. More generally, I would always suggest following the advice given in later sections of *Chapter 3, Safely Making Changes*. Just to reiterate, upgrades are always more complex and difficult than they first appear, and they need to be approached in a structured way.

Migration

What is it?

This is like upgrading, except the components are moved to a new technology or provider rather than later versions of the current technology. This can be applied to any part of the system across both the hardware and software stack. This may include:

- Operating systems (for example, Solaris -> Linux)
- Vendors of Java/CLR (for example, Sun JDK -> JRocket JDK)
- Database vendors (for example, Oracle -> Sybase)
- Processor architecture (for example, Intel -> Sparc)
- Communications infrastructure (for example, Cisco -> Juniper)
- Third-party products such as CRM/BPM systems (for example, Microsoft CRM -> Salesforce) and so forth.

Advantages

There are many reasons why an organization might choose to migrate rather than upgrade (or just maintain). For example, another vendor or product now has a better technical solution, or the original vendor may have ceased trading entirely. With legacy systems, it is very common for better technologies to have emerged and technology teams will want to take advantage of this.

There could be significant cost savings to be found by moving to a different product or consolidating to a single product across an organization. A good example can be found in databases where a site-wide license is purchased. There will be huge pressure to use this single product across the whole organization and stop paying maintenance fees for others.

Disadvantages

Migrations may be driven by technical or business requirements. It can be very frustrating for technical teams if the reason for migrating is cost, and equally annoying for the business owner if migrations are driven by technical desires. These kinds of conflicts are political risks and are dangerous to a project.

Many of the same issues I mentioned with upgrades are relevant to migrations but are often more severe. Problems with data aren't just due to bad data but also incompatible formats and missing, required data. You often need to enrich the data (make stuff up) to get a migration to work.

Like an upgrade, a migration may also necessitate modifications to components around a software/hardware stack.

Both upgrades and migrations will require operational changes in processes and in actions performed by the end user. The training overhead is likely to be greater for migrations than upgrades.

Advice on Implementation

There can be other drivers for the selection of technologies or requirements for migrations, such as finding suitable skills in the job market or support contract simplification. It is important for you to find out exactly why decisions are made.

The effort involved in migrating data between systems should not be underestimated. Although modifying formats can be challenging, it is missing data that will cause the most issues. You really need to make sure that these items are identified early and that you get agreement from the other stakeholders as to the resolution. It may not be possible to fill in missing data and you should make sure that you have a record of exactly what is 'made up' in order to get the system running. For example, if a new system insists that a phone number is included (and you don't have access to these) you should make sure that any 'fake' number you insert is obvious.

Lastly, please don't forget to modify any documentation to bring it into line with your upgraded/migrated system. It's very frustrating to think you have some relevant documentation only to find out that it refers to a much older version. It makes the user unsure as to whether it can be trusted or is useful.

Incremental Improvements

What is it?

In this strategy, you keep the basic infrastructure and system architecture the same (probably with some upgrades or migrations first) and then either add new components or add functionality to current components. Functional additions may be driven by internal requirements to improve or expand the product or external factors such as regulation. The changes could also be driven by non-functional requirements such as coping with an increased number of users (there is a debate about what separates a functional and non-functional requirement, which we won't go into here!).

legacy systems are frequently systems that have had several incremental improvements made over their lifetime. It can be amazing to try to track down the earliest unchanged source code file, the earliest date in a comment, or the most ancient piece of hardware in a system that's been changed beyond all recognition (I was recently told that a certain modern, 3D, popular football game still had all the 2D sprites in it from a version 10 years earlier that no one had ever gotten around to removing).

Advantages

Incremental improvement allows you to give end users specific functionality on a per-requirement basis. They can see a real, defined benefit rather than lots of background improvements that make little difference to their jobs. Hopefully, you can deliver this in regular, small deployments.

Disadvantages

There is often huge pressure from business owners to "just stick it in" and get the required functionality as soon as possible (are these deadlines for an external reason or just made up?) but this leads to the dreaded technical debt. It's important to refactor but **NOT JUST THE CODE BASE**. If the usage changes considerably, then you might also need to change the way that software components are hosted, run, and communicate. You need to apply refactoring techniques to the frameworks and infrastructure right down to the hardware. If you are starting with hardware changes, then this might also work up through the stack as well.

Advice on Implementation

I would strongly advise you to perform any upgrades required to bring the components to their latest versions before adding any functionality. This may be opposed by the business owner if incremental improvement is viewed as a cheap option ("just stick it in"). This may be particularly true if this is driven by external factors, such as regulation, where the organization won't see any tangible benefit.

If you are writing code to add features or fix old bugs, I suggest first creating a new baseline for the code. This involves getting all the files and simply formatting them and organizing them in your preferred project structure. The new baseline should not have any functional modifications and the files should be checked into source control and labelled.

This means that any functional modifications you make from this point will show as clean and simple diffs. Without a new/clean baseline, you will find that any diff you run will include formatting and structural changes (such as file moves). You should also perform this baseline formatting on configuration files such as XML. It is amazing how inconsistent the formatting of files can become over a period, especially with many people working on them, but also how formatting fashions change.

Replacement

What is it?

This is a complete re-write with no reuse beyond business knowledge. It is likely that some of the original system will be reused and, certainly, the data will be migrated. However, the intention is to replace as much as possible with a top-down approach that is, not refactoring and rewriting the system from the code upward but re-implementing from requirements down.

Advantages

This is often the preferred option for the technology team, as this gives maximum scope to use new technologies and techniques. It also avoids having to learn the idiosyncrasies of the legacy system and understand supplanted technologies. It also allows them to use familiar languages, tools, and equipment.

Disadvantages

However, you must ask yourself if you really understand everything the legacy system does or can do. We should remember the problems of lost, hidden, and implicit knowledge and understand that a replacement is very tempting but often incredibly hard to achieve. Do you really understand the business requirements or is this driven by the technology team's desires? All the issues listed for data in the *Migration* section are also true for a replacement.

The legacy system will have to be maintained while a replacement is being developed and new features might even have to be added due to external drivers. I have seen legacy systems that have had so much incremental improvement while a replacement was being developed, that the replacement never went live – it was constantly chasing a moving target.

Other costs to consider are those for the complete retraining of users and operations staff.

Advice on Implementation

The biggest and most common mistake people make when replacing legacy systems is to not understand all the functionality they are trying to replace. This is often because they assume the legacy system has little or no value, but this is a mistake – the system is legacy because it has value and you need to understand what this is.

You should pay attention to the data in the old system, as you will almost certainly need to import this into your new system. You don't want to get to the release stage of your replacement and realize that there is a large dataset that can't be imported or is not dealt with.

You should consider trying to run a replacement system and its legacy system in parallel configuration rather than a big-bang release. I'll go into this in more detail in the next section.

A Special Note on Decommissioning

Decommissioning is the process of shutting down a legacy system. This is often combined with a 'replacement' project/strategy, as once the new system is released, then the legacy system should be turned off. This sounds simple but often isn't.

Issues

You will not want to run multiple, overlapping systems of different ages and technologies but that is a very common outcome. Often, a replacement system will not cover all the functionality of the legacy system it supposedly replaces, and the legacy system will be left running to perform a small subset of this original functionality. This means the organization has all the maintenance issues and costs of the legacy system, as well as those of the new system (a friend recently commented that he has *never* known a system to be fully decommissioned and his organization was filled with almost dead, zombie systems). This includes multiple teams with overlapping responsibilities, and this invariably leads to complex politics.

Advice on Implementation

You need a specific plan for decommissioning – do not just assume you can turn it off. I would suggest, at a minimum, the following steps:

Firstly, you must make sure all the stakeholders are committed to the decommissioning – there may be many hidden agendas. Please refer to the *Stakeholders* section.

Secondly, you need to make sure you understand all the external connections and dependencies. You are interested in external systems that are tightly coupled with your legacy system. These may need special treatment for compatibility. Please refer to the *Architectural Sketches* and *Further Analysis* sections.

Lastly, you need to decide on your actual strategy for moving from legacy to replacement systems. This may include:

- The **Big Bang**

  This involves developing and then deploying an entirely new system and turning off the old one at the same time. This has the advantage of making sure the legacy system is deactivated. However, if your analysis missed some important functionality or dependency, you might find yourself having to turn the entire system back on again. This might get repeated for multiple features until you run out of development budget – at which point, you are left with both systems forever.

  You also must make sure that all users are completely trained for the new system on day one. This can be difficult and if you are forced into a rollback, there could be chaos.

- The **Parallel Run**

  This is similar to the Big Bang, in that you have developed (or possibly bought in) a complete replacement system, but you deliberately run both systems for a period of time and gradually move users and functionality from one system to the other in small increments. This has advantages in terms of user training and impact and means that a problematic feature can be individually dealt with rather than a large and embarrassing rollback.

  However, it's possible (due to time or budget constraints) to not fully move the system over or miss important features. Either will force you to leave the entire legacy system running, even though only a small subsection is now being used.

- **Agile Component-Based Replacement**

  Rather than creating and deploying an entire system, the individual components within it are created, deployed and decommissioned on a one-by-one basis. This reduces any single impact and means that any unknown functionality is handled in the same way as any other known functionality. This can be unpopular with project managers, who demand a fixed timescale and cost estimate, but is more likely to deliver the functionality that users want.

It can work out much cheaper than a Big Bang release, as only the functionality that is used is replicated. You simply need to select components/functionality, replicate it in your new framework, and repeat until there is nothing left. If there is a large amount of unused functionality, you will save yourself from unnecessary re-implementation. You need to be careful of hidden or lost knowledge and track the project carefully to make sure that the legacy system is deactivated at the end. The problem here is defining what the end point is.

Conclusion

You might (and probably should) mix and match the suggested strategies somewhat. It is also possible to treat each component in a different way, that is, maintain some, upgrade those that need it, and replace ones that need completely new behavior. You need to understand the business' motivation and set expectations accordingly for timescales and cost. As developers, we usually favor re-writing codes –users often want little to no impact (they have jobs to get on with) and the organization's management care about cost (and, very often, the focus is on short-to-medium term costs). Choosing a strategy is difficult and involves many trade-offs.

2

Investigation and System Review

Architectural Description and Review

When designing the architecture for a Greenfield system, you periodically perform architectural reviews to determine whether it can deliver the user's requirements. For a legacy system, you know it works (even if it has issues) but you need to determine how it works. Even with documentation, there is likely to be a gap between that and reality. You start with two extra important pieces of information, which are the real stakeholders and artefacts of the running system itself.

The International Standard for "Systems and software engineering – Architecture Description" is ISO42010. I will be using the terminology and processes from this standard where possible. Don't let this panic you! This is a lightweight standard and imposes no restrictions on your processes or development methodology. When creating a Greenfield Architectural Description, you will (usually) perform the following steps:

- Identify stakeholders
- Identify stakeholder's concerns
- Identify functionality and quality attributes to meet concerns
- Construct a model
- Create views – applying viewpoints to frame concerns
- Use views to code and construct the architecture/system

(These are deliberately not numbered as actions can be iterative or performed in blocks.)

You won't perform the same actions when reviewing a legacy system (and retrofitting/ writing an Architectural Description). The last item on the list – construct the system – has been done, so the temptation is to work backwards from that point, that is:

1. Create views on the current system (maybe using reverse engineering tools)
2. Construct a model based on these views
3. Identify the desired system functionality from the model/views
4. Identify stakeholders affected by the functionality

This may work on a small system, but caution is advised for several reasons:

1. Creating an architectural view from code and configuration is very difficult. Most code is not written in an **Architecturally Evident Style** (Refer–http://www. codingthearchitecture.com/2014/06/01/an_architecturally_evident_coding_ style.html for a discussion on this) and reverse engineering tools often miss important correspondences or show everything as connected.

2. Trying to determine the desired functionality from a view and model is very difficult (and quality attributes, even more so). You cannot tell the difference between a feature and a side effect. You will waste time discovering features that are not used.

3. You cannot determine who is using a feature or how it is used by examining the feature. You have no idea whether the tool is being used in a sensible or predictable way.

I would advise applying a variation of the 'normal' forward architecting (yes, 'architecting' is the correct term, as defined in ISO42010; even though no spell checker recognizes it) process, as well as reverse engineering the system:

1. Identify stakeholders (real users of the system)

2. Identify used functionality based on stakeholder use

3. Identify other concerns (issues with the system, required quality metrics, and so on)

4. Examine the system from a macro perspective downward (identifying architectural elements as you go)

5. Create lightweight views by sketching the system (avoiding trying to capture too much information)

6. Verify your sketches using the stakeholders' use cases

The following sections cover stakeholders, architectural sketches, and then a selection of analysis to gain a deeper understanding of the system and its runtime operation.

> **Note**
>
> In the following sections, I will be giving some examples based upon a simple scenario shown in *Appendix 1*. Please read this (it's a single page) so the examples make sense. These are illustrative and simplified and you should go into the depth required for you own system.

Stakeholders

Locate the stakeholders

It is important to identify as many of the stakeholders as possible. Too often, technologists identify a single 'user' of the system, who is helping drive the project, and only speak to them. You should locate all the people that can provide you with information *and* those that have other interests in the system. This not only helps you gather the information you require but also keeps people informed – which helps avoid corrosive politics later in the maintenance life cycle.

One of the benefits of working on a legacy system is that you have real users you can speak to rather than having to make assumptions.

> **Note**
>
> To understand a system, you need to understand its stakeholders and their concerns

Below are some suggested stakeholders and job functions for a system. Your system of interest may use different titles (I'm using fewer formal definitions than ITIL) so please feel free to use your preferred names.

I've included a short section called *How can you help?* to provide hints about winning a stakeholder's trust and assistance but there are two things you should do for every group:

- Listen to their concerns
- Keep them informed

When I say 'concerns' I'm referring to both the ISO42010 definition of Concern (an interest or requirement in the system from their point of view) but also their personal concerns (which might be largely emotional) about how the changes will affect them. People want to be listened to; remember that rumor and nonsense tend to fill an information vacuum.

> **Note**
>
> "Software Systems Architecture" (by Eoin Woods and Nick Rozanski) gives important advice on performing architectural reviews including "Don't call anyone's baby ugly." This is even better advice for sketching and reviewing legacy systems. Be careful to avoid making harsh criticisms when discussing a system with someone who helped create it!

End Users

Who are they?

These are the external users of the system (not part of the development or support team) using its main functionality. They may work for the same organization or be external customers. This is an obvious but diverse group so you should try to speak to a range of users, including those that:

- Create data

- Consume data

- Generate reports (logic applied across the data)

In our warehouse management scenario, this could include:

- Product designers (data creator) – who enter the details of products that can be stored and sold

- Salespeople (data creator) – who enter the details of orders that need to be fulfilled

- Warehouse manager (data consumer and report generator) – who receives data about what needs to be moved in and out of the warehouse based upon products and orders

- Forklift driver (indirect data consumer) – who is handed a daily work schedule generated by the system via the warehouse manager

These people have very different requirements and knowledge of the system. I've also included an indirect user (forklift driver) whose daily work is greatly affected by the system but isn't the person who interacts with it. This type of 'user' is often ignored but if they can't work efficiently, the entire system is pointless.

Although we often focus on these end users, they may not spend much of their working time using the system. In the preceding example, the system is core to their jobs and it is used frequently but those interactions are short. This is important but often forgotten by those of us that spend all day in front of a screen. This is even more likely to be true of external users.

What information can they provide?

These are the most obvious stakeholders and they can:

- Provide information about how they interact with the system
- Describe how the system fits into the workflow
- Tell you what the system does well and where it can be improved
- Point out bugs in functionality and non-functional issues they face

They can either be the loudest cheerleaders for your project or the people who most want to see it fail (see *Politics* section).

Concerns they may have

Following is the list of a few concerns:

- Losing functionality, they rely upon
- The system becoming more complex and difficult to use

How can you help?

I will reiterate how important contact and feedback is to reduce the fears they have. This is a varied group of people so be careful to treat them appropriately – don't use technical jargon if it won't be understood, as you'll only make them more concerned.

Try to identify important functionality for each user and let them know you understand how it's used and why it's important.

Support Staff

Who are they?

Anyone who directly helps the end users of the system; they may provide advice on use or perform administrative functions:

- Administrators (such as adding users)
- Moderators
- 'Super' users
- First-line phone support for external users
- Internal technical support
- nth line technical support

Some of the support staff may also be users of the system. (It's amazing how some commercial organizations can persuade their customers to become unpaid support staff by calling them 'moderators' or 'super users.') Therefore, it's important to identify the roles people have rather than just identify who they are. In our warehouse example, some examples might be:

- Warehouse manager (administrator) – who can add new warehouse staff to the system, such as forklift drivers.

- 'Dave in I.T.' (internal/external support) – who used to work in the warehouse and knows how to use the system as well as how it runs.

It's also worth remembering that not all of the roles will be official. In the preceding example, 'Dave' might have a small official role with respect to the system but due to his history, this is the first (unofficial) point of call. This is incredibly common with legacy systems as people change responsibilities and move jobs and departments; systems are outliving their user's jobs.

In this small system example, the support staff will spend minimal time performing administration tasks. However, for a large system (such as a logistics product used in a supermarket chain), the entire job of many support staff is based only on that system.

What information can they provide?

They will have a lot of information about issues with the system, including:

- Functional bugs

- Issues with the quality of service, for both them and the users

- Issues with usability, where functionality is not obvious to new or infrequent users

They will also have suggestions for making the system easier to administer and more transparent. The support staff may have users ranting at them when there are problems, and this may get passed onto you – try to not take this personally.

Concerns they may have

Following are some of the concerns:

- The system becoming more complex and increasing their workload through increased user issues

- Having to re-train – both themselves and the users they support

- Losing their jobs and 'change' (see *Hidden Knowledge* sub section in *Common Issues* section)

How can you help?

Listen to their problems with the current system and provide some solutions to some simple problems. There are bound to be some time-consuming, repetitive tasks they are forced to perform that you might be able to help with.

System's Infrastructure Staff

Who are they?

These are the staff that look after the infrastructure that the system relies upon. These can include:

- Database administrators
- Email/exchange administrators
- CRM/ERP administrators
- Network support
- Hardware support

I won't list the people in the example scenario, as we'll assume there is one for each of these roles and that they work for a central IT department. However, all the preceding are required.

We should remember to include staff from required external systems – from payment gateways to real-time data providers. They have information relating to your system's behavior and how you can improve its interactions. In our example scenario, this would be the external system controlled by the manufacturer.

What information can they provide?

The systems staff can provide a lot of information about the non-functional behavior of the system; from its IO behavior to CPU and memory requirements. They can often give you a better idea of when and how much the system is being used than can be provided from the system's own internal, monitoring tools. People from a software background are often unaware of the level of information available from the infrastructure teams. Some of their information can lead to great performance gains – most obviously from database administrators (who can give advice on indexes and cache sizing) but even suggestions on using email gateways correctly can help remove blocks to your processes.

The systems staff for external dependencies may also have similar advice. From when the legacy system was written to the present, the external systems might have improved considerably and have features you can leverage.

Concerns they may have

- Changes to the system having an impact on the infrastructure; for example, a new application server requiring more memory

- Changes to the use of the system impacting the infrastructure; for example, extra report runs loading the database

- Upgrades to the legacy system requiring major changes to their infrastructure; for example, a database upgrade

- Releases requiring weekend and evening work for supporting infrastructure

- Problems with a fragile legacy system being 'blamed' on underlying infrastructure

How can you help?

Investigate the monitoring tools the systems team uses and make sure you don't break the current monitoring. They may be monitoring a range of indicators, from the operating system through to suspect messages in log files. These could have been built up over several years, might not be obvious (refer to *Forgotten Knowledge* section, and be easy to silently break. If they are grepping log files, you must be careful.

Audit and Compliance

Who are they?

The individual (or department) responsible for checking and reporting compliance with regulations and industry standards. They may also be under a legal duty to report breaches to the authorities.

The requirement for audit and compliance will depend on the type of system and varies greatly between systems. However, even if you initially think there are no compliance requirements, you should still investigate and find out who is the responsible party. If no one knows who the responsible party is, then there's a good chance it's you!

As already stated in the *Common Issues* section, there is a problem with regulations changing and legacy systems not keeping up. Remember that non-compliance, IT security issues, or privacy breaches can have legal ramifications. You want to establish responsibilities *before* this occurs.

What information can they provide?

- Information about what needs to be externally reported

- New regulations that a system now needs to comply with

If the cost of modification is beyond the budget holders' initial expectations, there's nothing more likely to help get extra budget than mentioning the chances of prosecution.

Concerns they may have

- Modifications to the system will break compliance

- Your investigations into the system may show that they haven't been compliant for a while, causing them to compile a report and answer difficult questions

- Small functional changes can require a huge amount of work for re-certification

How can you help?

Make sure that you check with compliance about new features being added to the system or functionality they may not be aware of.

You should engage them at the very beginning of your project to find out what areas are of concern (for example, a financial project may have very different regulations to a factory automation project).

Budget Holders

Who are they?

The individual or department that controls the budget for your legacy system. They will be paying for the work and other items such as new licenses, and so on. In a small organization, these can be the users of the system but in a large organization this is rarely the case.

They may be powerful but ignorant as to the workings and importance of the system (which might pre-date all the individuals). You should try to engage these people early, get them on side, and keep them informed so that there are no nasty surprises.

What information can they provide?

The real motivation behind decisions! They may also have information about licenses and other assets associated with the project. You might have assets that are useful that you don't know about. For example, some of the hardware/software may have extended support or professional service time. Expensive support contracts may include this but not be being used. Using up your hours of expert support on third-party products can be lifesaving.

Concerns they may have

- Conflicts between projects with overlapping responsibilities.

- Being over budget, causing them difficulties in funding the project or cancelling it.

- Being under budget. Many organizations have a 'use it or lose it' policy for funding.

Weirdly, coming in exactly on budget is preferred by many budgets' holders to underspending. It demonstrates how good their planning and estimation is and means they won't have money removed from next year's budget.

How can you help?

Keep them informed. A good budget holder or project manager should be able to cope with some changes but the sooner they know, the better. Remember you must balance this against pestering them with every concern about the system. Remember that:

- They are *not* interested in minor technical issues. It's your job to sort them out.

- They *are* interested in anything that affects costs, from licenses to man-hours worked.

- They *are* interested in anything that affects timescales, such as overruns or external dependencies taking longer than expected.

Third-Party Suppliers

Who are they?

These are the organizations that supply the hardware and software used in the target system. There are likely to be multiple suppliers, even on a system written in-house (and more than one even if the system has one main supplier). For example, you should consider hardware suppliers, operating systems, virtualization vendors, application servers (and other containers), library authors, and so on. The days of a mainframe vendor supplying the entire stack are largely ancient history.

What information can they provide?

If the core of your system was supplied by a single vendor (with little modification), then they should be able to supply a huge range of useful information and professional services. Other vendors should also be able to provide supporting information. You will be interested in upgrade paths and compatibility.

If the system is old, then the supplier may have changed considerably and be able to offer less help. Some of the products you use might be discontinued. As mentioned elsewhere, you should discover whether the components supplied are still covered by support contracts.

Concerns they may have

- Their ability to support an old product; possibly to the point of not being able to meet old license terms

- Incompatibilities between upgraded products (either their own or the infrastructure underlying it)

- Old support contracts giving the client the ability to use services at below-cost

- Reputation issues relating to the above

> **Note**
>
> I really have come across a case where an old contract gave a customer the right to use a vendor's professional services for below-cost. This, of course, led to a very poor level of service being provided.

How can you help?

Try to build a relationship with the vendor and understand the difficulties they may have in supporting the product. You may have to let them support you on a 'best efforts' basis – which is not perfect but preferable to facing a wall of legalese if they try to avoid giving any support.

You should also be prepared to supply them with information about their own products and how they fit into your system. Don't assume they have adequate records or documentation any more than you do (for a fascinating story, called "Institutional memory and reverse smuggling"). You should use your context and containers diagrams to help facilitate your discussions.

Developers/Software Engineers/Database Analysts

Who are they?

The engineers that either helped develop the original system or those that will perform modifications to it. These may be the same or different individuals.

What information can they provide?

- Low-level implementation details

- Motivation, assumptions, and context for decisions

- Information about unused or difficult-to-find features

Concerns they may have

Opening a can of worms! They may be concerned about being asked to provide unofficial support for a product that they have no time allocated for. They may also have concerns about working with old technology again, which might not look good on their CV/resume.

How can you help?

Speak to their line manager and try to get some official time for them to do the work. Offer them a chance to pick replacement technologies that may help fulfil their career path.

Document the Stakeholders

Simple Stakeholder Context Diagram

I'm not suggesting you create a heavyweight document but take notes about the stakeholders you have and create a diagram that puts them into context with each other and the system. This diagram can be a UML use-case diagram (or set of them) or a very simple diagram like this example:

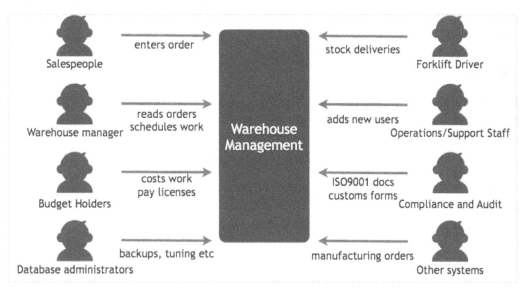

Figure 2.1: Simple stakeholders' diagram

A diagram like this is a useful tool to understand the conversions about the system. When shown to already identified stakeholders, they'll often suggest other people and functions that have been missed.

Concern-Stakeholder Traceability Matrix

If many stakeholders and concerns are identified, it can be difficult to track how system changes affect them. You need to associate concerns back to stakeholders affected by system/architectural changes. The simplest way to do this is with a **Concern-Stakeholder** table or spreadsheet, as follows:

	Stakeholder X	Stakeholder Y	Stakeholder Z
Concern A	Yes	No	Yes
Concern B	No	No	Yes
Concern C	Yes	No	No

Figure 2.2: Example Concern-Stakeholder Table

I record this in a spreadsheet so I can filter on columns to find a stakeholder's concerns (when I have an imminent meeting) or highlight a row to know who to discuss a modification with. The 'concerns' listed in the matrix should include anything important to the stakeholders and not just negative issues. For example, it should list core functionality they use and are happy with – users can be unhappy with any changes, even improvements.

Hyperlinking the concerns to more in-depth documentation and related changes can be helpful. Once set up, this has low maintenance.

Architectural Sketches

Why Sketch?

I've already mentioned creating a simple stakeholder diagram as a tool to aid investigation and conversation. It is also important to create an architecture diagram for the same reasons. It is much easier to discuss how a system works and required changes if you can physically point to something and discuss it.

Do you really know what you have and where it is? Do you really know how it interacts with the outside world and with the infrastructure below it and the services on top of it? Legacy systems can be complex with a non-obvious structure.

There are an infinite number of ways to describe and document a system, from high-level overviews to low-level detailed descriptions of all components. You can show cross sections through any part of the system. The danger is that you will try to show *everything* on a single diagram and make it impossible to use as a guide to the system.

Architectural Elements

In Architectural Description terminology (As defined in ISO42010), anything that you can draw in a view to a construct of an architecture is referred to as an 'Architectural Element'. This is irrespective of the level of abstraction. So, anything from the entire system to an individual class is considered an element.

Every stakeholder of the system will have their own preferred view of the system (referred to as a Viewpoint). These might be very similar for a simple system but differ greatly for a complex one. For example, a database analyst views the system (and the components they interact with) differently from that of a data entry clerk. The data analyst will visualize the system as a set of entity relationships, whereas the data entry clerk will only see an end-user GUI. Ideally, you'd document a system with a set of views for each of the stakeholders that detailed everything they needed to know.

This is a time-consuming task, involves duplication, and is unlikely to be kept up-to-date. Therefore, most architects will draw diagrams of the system that break it down in a standard way. There are many published frameworks and methodologies for doing this, but these can be heavyweight, and are often designed for building complex, Greenfield applications over long timescales.

Isn't the code the point of truth?

The source code (or OS/hardware configuration) is a view into the system but certainly not the only one. It doesn't describe where and how it is deployed, and it is difficult to navigate at a high level. Most importantly, it doesn't help with conversations regarding the system with non-technical stakeholders.

This is where architectural sketches can really help. Simon Brown's book, *Software Architecture for Developers*—https://leanpub.com/software-architecture-for-developers, is a great resource and guide to doing this. Simon also runs a two-day workshop where this is taught in more detail.

I won't repeat the material here, but I will give simple examples applied to our example legacy system (refer *Appendix 1*). If you require more details, then please download one of the many free essays on the subject at http://codingthearchitecture.com

Context Diagram

What is it?

A context diagram shows where and how the system relates to the world around it. It should answer the following questions:

- What is the system?

- Who is using it?

- How does it fit in with the existing IT environment?

What does it contain?

At its most basic, it is a simple block diagram showing your system as a box, along with its users and all the other systems that it interfaces with. In other words, its dependencies. You are likely to include many of the stakeholders identified in the previous section.

Why is it useful?

It's a starting point for discussion with all the stakeholders. It allows you to very quickly show where you believe they interact with the system and talk about how they interact. I *guarantee* that when you initially draw this diagram, and discuss it with your stakeholders, they will point out extra interactions that you've missed. It's much easier to discuss a diagram than a block of text.

It also provides a starting point for identifying who you need to go and talk to as far as understanding inter-system interfaces are concerned. Again, it is very likely that, once you start these discussions, you will discover more dependencies.

Identifying Context Elements

These course-grained elements should be easy to identify from conversations with stakeholders – of course, the main one is the system you are examining! However, it is possible that some external systems will not be immediately obvious. Examining network traffic and firewall rules may identify additional external dependencies. Make sure you show your 'complete' context diagram to network administrators and check whether anything is missing from a communications perspective – you may be surprised.

A simple example

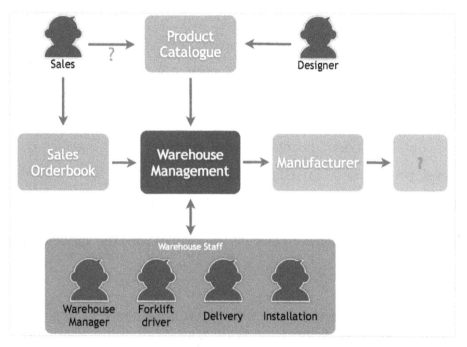

Figure 2.3: Simple context diagram for warehouse management

The preceding example is very simple, and you would want to include a little more detail. For example, it would certainly benefit from a key – what do the different colored boxes mean?

This is a legacy system, so in theory it should be possible to sketch all the elements. The problem of forgotten/hidden knowledge means that you might not be able to fill everything in at the beginning. Please do add indications of what is unknown. In my example, I'm not sure whether the sales team is editing product information (imagine a situation where someone in the sales team is tweaking product information to make it more desirable to the customers they are selling it to; they might be a little reticent in admitting it) and I suspect that the manufacturer is forwarding orders to subcontractors.

Remember, this is *your* diagram, so add what is useful but avoid making it too cluttered.

Container Diagram

What is it?

The purpose of the container diagram is to illustrate the high-level technology choices. It should answer the following questions:

- What are the high-level technologies used?
- How do containers communicate with one another?
- As a developer, where do I need to write/maintain code?
- As an operations/system team member, what do I monitor?

This is a diagram that you will use to discuss the system with the various technical teams.

What does it contain?

The logical executable or processes that make up your software system, for example:

- Web servers (for example, Apache Tomcat, Microsoft IIS, WEBrick, and so on)
- Application servers (for example, IBM WebSphere, BEA/Oracle WebLogic, JBoss AS, and so on)
- Enterprise service buses and business process orchestration engines (for example, Oracle Fusion middleware, and so on)
- SQL databases (for example, Oracle, Sybase, Microsoft SQL Server, MySQL, PostgreSQL, and so on)
- NoSQL databases (for example, MongoDB, CouchDB, RavenDB, Redis, Neo4j, and so on)
- Other storage systems (for example, Amazon S3, and so on)
- Web browser, plugins and so on

For each container drawn on the diagram, you could specify:

- Name

- Technology

- Responsibilities

- Interactions – communication protocol and style

Again, like the context diagram, you can put as much detail in as you like but don't try to document the entire system on a single diagram.

Why is it useful?

In a large organization, you may have several technical teams that need to work together on your projects: application developers, database administrators, hardware support, network administration, email support, CRM, and so on. You need to be able to talk about where and how these runs and communicate and who supports them.

In a complex system, this diagram will be the first place to look when trying to identify system issues; for example, when you have a query from a user such as:

"My data X seems to be stale as it hasn't updated for 3 hours."

You can look at this diagram and work out where this should have come from and how it should have gotten from the point of generation to use.

Identifying Container Elements

As containers are logical executable or processes, these should be easy to identify by looking for software services running on known hardware. This means knowing what hardware is associated with the system and getting access to it. Your system operators should be able to help you with this.

You may also want to check asset registers for hardware and associated software licenses for what should be running. If you can examine your network traffic, then each endpoint of communication will be a service/container.

Build systems such as Ant, Maven, NuBuild, and so on each have hooks in the code base for building and deployment. An extraction of the build targets will give you the deployment modules. This may give you the required information for a **Containers View**.

A simple example

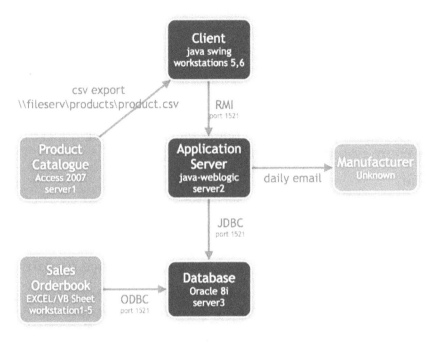

Figure 2.4: Simple container diagram for warehouse management

This brings up some interesting observations on the example system:

- Sales are pushed directly into the DB 'under the covers' via VBA calling ODBC.

- The Excel spreadsheet seems to be stored on PCs rather than centrally or in source control. Alarm bells are ringing!

- The product catalogue uses Access and is newer than the Java stack.

- The feeding systems don't call the server directly; they use an export/import of text files.

- The manufacturer gets the orders (for low-running stock) via a daily email. Simple and not prone to errors but has long latency. Is this a problem for users?

Your own sketches should help you identify these kinds of issues with your system.

Component Diagram

What is it?

The component diagram allows you to zoom into your containers and decompose them. This is about partitioning the functionality implemented by a system into several distinct components, services, subsystems, layers, workflows, and so on. You can do this with both the software and hardware and other details such as the networking.

Yet again, I'd warn against trying to show too much in any individual diagram. It probably makes sense to have at least one individual breakdown for each container and to keep the hardware and software breakdowns separate (although there are exceptions if parts of software and hardware are tightly coupled).

What does it contain?

The coarse-grained building blocks of your system. You should be able to understand how use cases, defined by your stakeholders, are implemented across these components. If you can do this, you've probably captured everything. For example, if you know that data entry is audited but you haven't sketched an audit component, you've missed something.

For each of the components, you could specify:

- The logical name of the component (for example, 'Product Importer', 'Inventory Reporter', and so on)

- The technology choice for the component (for example, Enterprise JavaBean, Windows Communication Foundation service, and so on)

- A very high-level statement of the component's responsibilities

Why is it useful?

These components are the blocks that you are most likely to deal with in your chosen strategy. For example, you may choose to upgrade and improve some components or replace others. This is the diagram that you can really have in-depth discussions with the teams about – any further breakdowns are likely to be performed by individual implementers.

Identifying Component Elements

Components are much harder to identify than context elements or containers, as they are not standalone items and are 'hidden' within the containers. They may be distinctly identifiable in a decoupled and cohesive system but almost impossible in a 'big ball of mud.' You might have to spend a reasonable amount of time scanning code and configurations in order to make some assumptions about what should be there. The following suggestions may help you track down your components:

- Source code packages/namespaces

  Many programming languages, such as Java and C#, group class files within packages/namespaces. If the system has been written using package-by-component (please refer—http://www.codingthearchitecture.com/2014/06/01/an_architecturally_evident_coding_style.html for a discussion on this topic), then the packages will relate one-to-one with the components. The links between components should be identifiable by the links between the classes within them (the has-a relationships). If the code is package-by-layer, then this is much harder to use. Experience tells me that real-world systems are a combination of the two, so there should be useful information but will require some careful examination.

- Class names

  It's very common for class names to contain a strong indication of their role; for example, XyzService, XyzConnector, XyzDao, XyzFacade, and so on. Scanning for known patterns should identify the element names and roles.

- XML DI configuration files

  Many projects use XML dependency injection files to define components and 'wire' them together. These show a relatively low-level model and you will need to do some manual tweaking to identify components. The properties used in the 'wiring' indicate the connections between these component elements.

- Interfaces and class hierarchies

  Certain interfaces (or extended base classes) may identify components and types; for example, implementing service, DAO, connector, repository, and so on.

- Module bundling systems

  Modular systems such as OSGi define bundles of components and services including life cycle and service registry. The deployment information should provide a high-level overview.

- Delegation or library dependency

  Shared delegates used by a set of classes/functions may indicate their purpose; for example, components delegating to a database utility might indicate a DAO component, or using a CORBA utility might indicate a service. Finding these will be time-consuming, as you need to identify and scan for each delegate you are interested in.

- Comments/Javadoc/JSDoc/NDoc/Doclets

  Comments and Javadoc style API comments can provide a large amount of meta-information about a class or package. In fact, many UML modelling tools enrich code using custom comments and tags.

- Tests

  Tests can provide a lot of metadata about your system. Unit tests tend to be concentrated around important classes and often construct entire components to test. Simply extracting the names of classes that are directly tested will produce a useful list of components. Higher-level systems tests will reveal important services.

- Annotations

  Annotations are present in Java EE applications and other enterprise frameworks and can quickly identify components.

- Communication frameworks

  The configuration for communication frameworks (such as named, asynchronous, and messaging queues) contain information about how the system is decoupled and about course-grained components.

A simple example

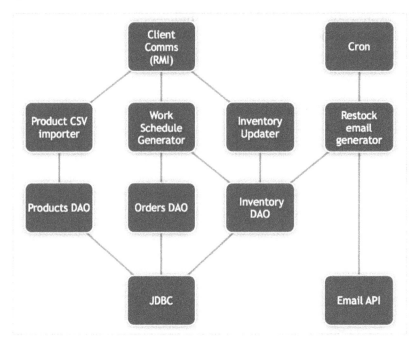

Figure 2.5: Components in the Warehouse Management App in the Application Server container

The first thing to note is that you probably don't like the 'design,' but this is what exists in reality – you're showing what's there rather than what you'd like to see. For brevity, I've not included the technologies in the diagram or responsibility statement.

In order to determine whether you have captured the important components and their connections, you should walk through the important use cases identified by your stakeholders:

- Daily email to manufacturer:

    The cron job kicks off at 5.30p.m. and calls the restock email generator. The restock component calls the inventory DAO to get a list of products that have fallen below a certain level. An HTML email is formatted with details of the products required. An email is sent to the manufacturer with the products to be made.

- Designer imports new products to the system:

  They will have already exported the CSV list from the external product management system, which they load into the client (external). The client calls into the application server via the RMI client comms module with the CSV as a string message. Client comms passes this to the product importer. The product importer tokenizes the data and pushes it into the database. Success or failure is propagated back.

- Warehouse manager generates loading schedule for the day:

  The work schedule generator receives a request via client. The generator retrieves a list of orders to be fulfilled today from the order DAO. The generator retrieves an inventory from the inventory DAO. The generator calculates the most efficient way to fulfil orders. The generator creates a list of work items and returns a report.

- Warehouse manager updates inventory after a stocktake.

Use cases that do not involve the container/components directly would not be mentioned or checked here. For example, the orders are written straight to the database and product editing is handled elsewhere.

Those of you that use UML will have noticed that you can easily produce sequence diagrams at this point. However, this is probably not needed, unless it is very complex or you're about to perform a major refactor.

> ### Scenario-Based Evaluation
>
> Applying your known use cases to the Context, Container, and Component diagrams in this way is a simple type-of 'Scenario-Based Evaluation'. There are some formal (and heavyweight) techniques, such as **ATAM (Architecture Tradeoff Analysis Method)** and **SAAM (Software Architecture Assessment Method)**, which provide a detailed framework to do this. "Software Systems Architecture" by Eoin Woods and Nick Rozanski contains an accessible overview of ATAM.

C4 – Classes

What is it?

A breakdown of the actual objects used in the implementation. This will be classes/procedures/functions for the software, configuration files (for items such as the OS or database) and wirings/configuration for networks and so on. This level of breakdown is unnecessary unless you want to perform detailed changes to the component in question.

You're best off leaving this until you must do it, so I'm going to skip giving any more information. I can recommend the excellent book *Working Effectively with Legacy Code* by Michael Feathers.

Further Analysis

What Else Can Be Done?

The analyses suggested so far have been static and high-level. What follows are a few lower-level analysis areas that may apply to your system. Each of these short sections could justify their own chapters (or even whole books) but I will just give a few pointers for brevity.

If I've missed out anything you consider important, please email and let me know – perhaps you're working on a type of system I've not come across before.

Further Static Analysis

Database Schemas

There will be a data model (object/class model or data structure) within the application code and probably a separate one within a database. It is very common for these to be out of sync or for one to be a subset of the other. The database in question may be a traditional SQL one or something more unusual – currently, NoSQL databases are becoming popular but there is nothing new about changing trends and fashions in databases.

Fortunately, most databases follow a standard (SQL especially), which makes investigation easier, and they have a powerful set of tools to use. If you are lucky, you have a database with referential integrity and consistency checks. There may be views and stored procedures that give an insight into how the application uses the data store. Watch out for hidden logic in triggers!

Files

Older legacy systems (or ones that were compatible with the data of older systems) might store a lot of information in files. These files can be 'flat' (for example, .csv) or might have a complex data structure inside them, for example, xml or serialized objects).

They might follow a standard but are usually completely bespoke. This makes reverse engineering them very difficult unless you have example files and the code that both writes *and* reads them. This code might be spread around several different components. In the past, I've been forced to trap the file access, exercise the system, and work backward to the code that is using the file.

It's easy to break your ability to load data if files are serialized binary and you modify some code. It's also surprisingly easy to corrupt files or have issues due to something as simple as default line endings or text encoding. You might have to investigate the file format and the data model described within it.

It's tempting to ignore the application's files, but their structure is important in the same way, and for the same reasons, as a database schema.

Communication Protocols

Communication protocols are like file formats in their varied and frequently bespoke nature. However, unlike a file, there is also the dynamic and interactive nature of the communication state.

Normally, with a file, you'll have a single writer, which will complete its job, before a single reader accesses the data sometime later. When components are communicating, there may be a complex set of messages exchanged and a different message can completely change the conversation. The internal state of any of the communicating components will also change the contents of the communication. This makes any reverse engineering of a communication protocol very challenging. So, hopefully, you have some documentation.

Configuration

The saying "configuration is code" is very true. Configuration settings can dramatically change the functional and non-functional behavior of a system.

> **Note**
>
> Coincidentally, on the very same day I wrote the preceding paragraph, I was affected by database configuration that was set incorrectly. I was testing a system that had an upgraded database and was getting some strange results. It turned out that the DBAs had moved the data correctly but had failed to copy over all the settings. This meant that the default timestamp format was now different and (due to a badly written application) this was causing errors.

We tend to ignore configuration for parts of a system that we're not experts in but any configuration from any component in any layer is important. What isn't defined is important as well. It's very common for default values to change between versions of software, operating systems, and hardware. If you are performing upgrades to components, then the implicit settings from the defaults can easily catch you out (I started writing anecdotes about this, but they became too long and bitter).

Configuration changes should be treated in the same way as code changes; for example, stored in source control, versioned, and tested. For example, can you roll back to your previous database configuration? I would suggest that, as part of any initial investigation into a legacy system, you should document where the configuration is and make an initial copy. These might include:

- Database
- Network
- Storage
- Virtual machines
- Automatic backups
- Power-saving settings
- Email services
- Messaging systems

I bet you can think of many I've missed (please send them to me to add to the list!).

Source Code

There are an infinite number of tools for analyzing, storing, versioning, and building source code. I won't attempt to describe them. The source code should tell you a lot about the way the system works but be careful – what you have in the source control may differ from what is deployed. Check it against what is deployed and your binaries. You also need to check for nasty little presents such as well-hidden patch files (Java classpaths are a classic for this).

If you have a source control system with the complete development history of your system, then you are sitting on a goldmine – you just need to do some digging! The nuggets you find could include:

1. Check-in comments explaining decisions made

2. Lost configuration files and options

3. How configuration options relate to code (groups of files checked in together)

4. Removed tests that document system operations (very common when people used to write 'main' methods to test rather than xUnit)

5. Documentation that was considered redundant and removed

6. The identities of people who worked on the project (they may still be in the organization but have moved department)

7. Indications of problematic areas (where many changes have been made)

There is a lot of data and metadata available if you care to look. You might even be tempted to examine an early version of the system, before it became bloated with unnecessary features, to see how the core of it works (hat tip to Richard Jaques, who told me that if you want to understand how it works, just find the original version (that Linus wrote in a week or two) and read the source code... everything else is an extension).

Further Dynamic Analysis

Finally, we've finished the static analysis! Now we should analyze the dynamic behavior of our legacy system. This is important even if the current non-functional/quality behavior is acceptable, because you need to be able to tell if any quality attributes degrade after changes are made.

Some of the tools needed for this analysis may be present in your software or its platform but there are many fantastic tools your infrastructure team will know about (and if you don't have an infrastructure team, then you should find them yourself).

Stored Data Sizing

Having analyzed the structure of your data (databases, files, and so on), you should monitor it for growth. It may stay the same, gradually increase, or fluctuate on a daily/weekly basis (as data expands and is then culled via purging and clean-up tasks). I've never seen a system where the data usage shrinks. Although, I suppose that is possible.

It is very useful to know the profile of the data sizing over a period – I'd suggest a couple of months. This means that you can spot aberrant system behavior and is particularly useful after you make modifications to the system. It's very easy to make changes that cause unsustainable data growth and you need to know this before it causes a failure. I've seen a system sensibly improved but with a resultant side effect of generating much more data – when other, seemingly unrelated performance characteristics have been improved.

It should be easy to run database administration tools to find good statistics on database usage and if you don't have a DBA team, these are worth learning yourself.

Other data sizes and statistics can be more problematic. For example, if data is stored in a file, which grows and shrinks during usage, then you want to know the maximum size it reaches. Scanning for this data at the same time, once a day might not provide enough information. You may not even be fully aware of all the files a large application is using. Again, system tools can come to your aid here in both finding all the file handles used and individual file statistics.

Network Traffic

It's also useful to know the normal communications profile for your legacy system. In the same way that improved performance can increase the stored data produced, it can also increase the network traffic produced. It can be easy to flood a network or swamp downstream systems.

This should cover data consumption as well as data production. If you are using network tools to monitor this activity, then you'll see both as data quantity (and might not be able to distinguish between them). If you are using monitoring tools built into your applications, then you may only be seeing one side or the other, or perhaps part of the activity.

It is *very* useful to validate any data statistics generated by application tools with system tools; for example, comparing an Ethernet monitoring tool to your application or platform.

You may need to take this a stage further and gather some statistics about the type of message being sent as well as raw data quantities. This can be important if you're trying to protect other dependent systems (demanding huge quantities of data from upstream systems can have detrimental effects).

Process Activity

The normal operational processing profile (CPU/memory use, and so on) can be monitored at various levels throughout a system; from the application through the software platform to the operating system, virtualized machines, down to the hardware itself. These tools have improved dramatically in recent years and you might be surprised at how little visibility you have for a legacy system. As mentioned in the *Common Issues* sections, sometimes monitoring is lost or removed as non-core functionality.

It's very useful to have an operating profile for your system before any changes are made but this kind of monitoring can be intrusive (and therefore a change to the system in itself). These are often referred to as Heisenbugs. When you try to measure a problem, then the problem itself changes.

I would suggest adding this monitoring at the lowest level possible to start with and only adding it to higher levels if finer detail is required (that is, from the hardware up). This is because you are less likely to have functional side effects but it is contrary to most application developer's instincts, which are to add monitoring at the highest level.

Security Considerations

Information security is a quality attribute that can't easily be retrofitted. Concerns such as authorization, authentication, access, and data protection need to be defined early so they can influence the solution's design.

However, many aspects of information security aren't static. External security threats are constantly evolving, and the maintainers of a system need to keep up-to-date to analyze them. This may force change on an otherwise stable system.

Functional changes to a legacy system also need to be analyzed from a security standpoint. The initial design may have taken the security requirements into consideration (a quality attribute workshop is a good way to capture these) but are they reconsidered when features are added or changed? What if a sub-component is replaced or services are moved to a remote location? Is the analysis reperformed?

It can be tempting to view information security as a macho battle between evil, overseas (people always think they come from another country) hackers, and your own underpaid heroes, but many issues have simple roots. Many data breaches are not hacks but basic errors.

> **Note**
>
> I once worked at a company where an accountant intern accidentally emailed a spreadsheet with everyone's salary to the whole company.

Let's have a quick look at some of the issues that a long-running, line-of-business application might face.

Passwords

Passwords are every operations team's nightmare. Over the last 20 years the best practice advice for generating and storing passwords has changed dramatically. Users used to be advised to think of an unusual password and not write it down. However, it turns out that 'unusual' is actually very common, with people picking the same 'unusual' word. Leaked password lists from large websites have demonstrated how many users pick the same password. Therefore, the advice and allowable passwords for modern systems have changed (often, multiple-word sentences). Does your legacy system enforce this or is it filled with passwords from a brute-force list? (The LinkedIn website was hacked in 2012 and 165,00 passwords were easily extracted (http://mashable.com/2012/06/08/linkedin-stolen-passwords-list/)).

In-house systems are often implemented with a basic 'homegrown' username and password solution. These are often badly written, insecure (for example, storing passwords in clear text) and force the user to remember (and an operations team to support) yet another password. If this is the case, then please consider moving the system over to a **Single sign-on** (**SSO**) solution.

Passwords frequently get shared over time. What happens when someone goes on holiday, a weekly report needs to be run, but the template exists within a specific user's account? Often, they are phoned up and asked for their password. This may indicate a feature flaw in the product but is very common. There are many ways to improve this – from frequent password modifications to two factor authentications – but these increase the burden on the operations team.

Account Maintenance and Life cycle

Passwords fit into the wider issues of account maintenance and life cycle. Does your organization have an employee leaver's process, and do you suspend account access?

All systems have ways to create new accounts, so users can start using the system, but many have no real way to remove accounts. Often, this is required so system audits make sense (so actions performed can be linked to a user and this referential integrity is maintained), and in this case, the accounts should be suspended/locked for access. However, some systems are simply missing this essential feature. If this is the case, then I'd suggest setting the password to a long, random one and using this to make the account inaccessible. This is good practice anyway for accounts that are being suspended.

Of course, your system may allow for passwords to be reset by an administrator or accounts to be un-suspended. If this is a concern, you will have to introduce other operational procedures. There are several standards regarding this area, such as ITIL (http://en.wikipedia.org/wiki/ITIL#Information_security_management_system).

If you have shared accounts ("everyone knows the admin password"), this may be difficult or disruptive. Having a simple list (or preferably an automated script) to execute for each employee that leaves is important.

Cryptographic Keys

There are similar problems with cryptographic keys. Are they long enough to comply with the latest advice? Do they use a best practice algorithm or one with known issues? It is amazing how many websites use old, or even expired, certificates that should be replaced. How secure is your storage of these keys?

Are any of your passwords or keys embedded in system files? This may have seemed safe when the entire system was on a single machine in a secure location but if the system has been restructured, this may no longer be the case. For example, if some of the files have been moved to a shared or remote location, it might be possible for a non-authorized party to scan them.

Certificates

Certificates eventually expire and if they were issued with a long initial date, the creation process may no longer be understood. Renewals may be difficult if the keys or other details have been lost. Replacements may take time to be issued so it is important that the process is well understood, and renewal performed within a suitable timeframe.

Third-Party Passwords and Keys

Your system may interact with remote data providers and services. If this is the case, you will need to hold authentication information for those third parties. All the advice about storing keys, passwords, and renewal applies to these as well but you also have less control over them. Make sure that processes are well-documented and never store these in clear text. These third parties may have weaker policies than your own – you may need to push them to implement stronger passwords or cryptographic algorithms.

Lack of Patching

I've already mentioned patching several times; as a common issue and a challenge while making changes. When analyzing a legacy system, the following questions need to be answered:

- Have all the vendors' patches, related to security, been applied?
- What about the software stack beneath any application?
- Have any vendors applied patches to third-party libraries that they rely upon?
- What about the version of Java/.net that the application is running or the OS beneath that?

When an application is initially developed, it will use the latest versions, but unless a full dependency tree is recorded, the required upgrades can be difficult to track. It is even easy to forget these dependent upgrades on an actively developed system.

Even with a record of all components and subcomponents, there is no guarantee that, when upgraded, they will be compatible or work as before. The level of testing can be high, and this acts as a deterrent to change – you only need a single broken component for the entire system to be at risk.

Moving from Closed to Open Networks

A legacy system might have originally used a private, closed network for reasons of speed and reliability, but it may now be possible to meet those quality attributes on an open network with vastly reduced costs. However, if you move services from closed networks to open networks, you must reconsider the use of encryption on the connection. The security against eavesdropping/network sniffing was a fortunate side effect of the network being private, so the requirement may have not been captured – it was a given. This can be dangerous if the original requirements are used for restructuring. These *implicit* quality attributes are important and whether a feature change creates new quality attributes should be considered. You might find these cost-saving changes dropped on you by an excited accountant (who thinks their brilliance has just halved communication charges) with little warning!

Moving to an open network will make services reachable by unknown clients. This raises issues from Denial-of-Service attacks through to malicious clients attempting to use bad messages (such as SQL injection) to compromise a system. There are various techniques that can be applied at the network level to help here (VPNs, blocking unknown IPs, deep packet inspection, and so on) but ultimately the code being run in the services need to be security aware – this is very, very hard to do to an entire system after it is written.

Migrating to an SOA or micro services increases these effects, as the larger number of connections and endpoints then need to be secured. A well-modularized system may be easy to distribute but intra-process communication is much more secure than inter-process or inter-machine.

Modernizing Data Formats

Migrating from a closed, binary data format to an open one (XML) for messaging or storage makes navigating data easier, but this applies to casual scanning by an attacker as well. Relying on security by obscurity isn't a good idea (and this is not an excuse to avoid improving the readability of data) but many systems do. When improving data formats, you should reconsider where the data is being stored, what has access, and whether encryption is required.

Similar concerns should be addressed when making source code open source. Badly written code is now available for inspection and attack vectors can be examined. You should be careful to avoid leaking configuration into the source code if you are intending to make it open.

New Development and Copied Data

If new features are developed for a system that has been static for a while, it is likely that new developer, test, QA, and pre-production environments will be created. (The originals will either be out-of-date or not kept due to costs.) The quickest and most accurate way to create test environments is to clone production. This works well but copied data is as important as the original. Do you treat this copied data with the same security measures as production? If you have proprietary or confidential customer information, then it should be. Note that the definition of 'confidential' varies but you might be surprised at how broad some regulators make it. You may also be restricted in the information that you can move out of the country – is your development or QA team located overseas?

Remember, you are not just restricting access to your system but your data as well.

Server and Infrastructure Consolidation

Systems that pushed the boundaries of computing power 15 years ago can now be run on a cheap commodity server. Many organizations consolidate their systems on a regular basis, replacing multiple old servers with a single powerful one. An organization may have been through this process many times. If this has been done, you need to question whether this has increased the visibility of processes/services. If done correctly, with virtualization tools, then the virtual machines should still be isolated, but this is still worth checking. A more subtle problem can be caused by the removal of the infrastructure between services. There may no longer be routers or firewalls between the services (or virtual ones with a different setup) as they now sit on the same physical device. This means that a vulnerable, insecure server is less restricted – and therefore a more dangerous staging point if compromised.

A server consolidation process should, instead, be used as an opportunity to increase the security and isolation of services, as virtual firewalls are easy to create and monitoring can be improved.

Improved Backup Processes

Modifications to support processes can create security holes. For example, consider the daily backup of an application's data. The architect of a legacy system may have originally expected backups to be placed onto magnetic tapes and stored in a fire-safe near to the server itself (with periodic backups taken securely offsite).

A more modern process would use off-site, real-time replication. Many legacy systems have had their backup-to-tape processes replaced with backup-to-SAN, which is replicated off-site. This is simple to implement, faster, more reliable, and allows quicker restoration. However, who now has access to these backups? When a tape was placed in a fire-safe, the only people with access to the copied data were those with physical access to the safe. Now it can be accessed by anyone with read permission in any location the data is copied. Is this the same group of people as before? It is likely to be a much larger group (over a wide physical area) and could include those with borrowed passwords or those that have left the organization.

Any modifications to backup processes need to be analyzed from an information security perspective. This is not just for the initial backup location but anywhere else the data is copied to.

Checklist of Questions and Actions

Information security is an ongoing process that has multiple drivers, both internal and external to your system. The actions required will vary greatly between systems and depend on the system architecture, its business function and the environment it exists within. Any of these can change and affect security. Architectural thinking and awareness are central to providing this, and a good place to start is a diagram and a risk storming session. The following checklist is not intended to be comprehensive but gives a starting point for these discussions. The following taxonomy of operational cyber security risks and security risk framework may aid a more detailed discussion (http://resources.sei.cmu.edu/asset_files/TechnicalNote/2014_004_001_427329.pdf and http://resources.sei.cmu.edu/asset_files/TechnicalNote/2014_004_001_91026.pdf):

Question	Actions
Does the organization have an SSO solution?	Try to integrate if you can.
Is there a passwords policy?	If not, why not?! Implement one.
Does the password policy follow latest best practice?	Modernize the password policy. Expect some complaints though.
Have you ever audited the current password set?	Employ a security consultancy to test a brute force of your password list.
Do employees share users/passwords?	Find out why. Do permissions need updating or information shared using another method?
Does the system use any certificates?	Make sure the renewal process is well-documented and you have this planned well in advance of any expiry.
Does the system access third-party systems?	Perform an audit of the security provided by these systems as well as your own.
Does the system have a procedure for account suspension/removal?	If not, then introduce one. You may discover other weaknesses when you do this.
Does the organization have policies for employees leaving?	The account removal/suspension should fit into a wider process.

Figure 2.6: Checklist of questions and answers

Question	Actions
Can accounts be suspended?	If not, then you need a process to make them unusable (such as assigning a random password).
Have you checked your cryptographic key length/algorithm recently?	Best practice changes often. This should be reviewed regularly.
Are passwords/keys embedded in files? Where are they stored?	If the physical location has changed since the system was designed, you need to review from a security perspective.
Have sensitive files moved to shared drives/locations?	If the physical location has changed since the system was designed, you need to review from a security perspective.
Have all the vendors' patches, related to security, been applied?	Apply the patches! When testing patches, you should be careful about copying production data.
Are services running on an open network/internet?	Check if communication protocols are appropriate compared to a closed network.
Is duplicated production data used for test systems?	Check if this is compliant with data protection regulation in your industry and location. You may need to apply anonymization or treat the test system as you would production.
Have servers been consolidated?	Check there is no increased visibility between servers and their files.
Has infrastructure been consolidated?	Check the network access has not increased.
Where are backups stored? Has this changed?	Check who and what now has access to the backups. These need to be treated with the same care as production data.
What data protection regulation affects the system?	Your system may be old, but it still needs to comply with the latest regulation. This needs to be regularly reviewed.

Figure 2.7: Checklist of questions and answers continued

3

Making Changes

Safely Making Changes

If you are lucky, you work in an environment where it's acceptable to take risks, make changes, and sometimes break things. In my experience, this is rarely the case when dealing with a legacy system. The users are unlikely to be used to change and you need to goodwill with them before any negative impacts.

This section covers how to make changes to your legacy system in a predictable and safe way. Much of this should be useful even if the system is intended to be kept as functionally unmodified as possible.

Please note that the advice here is more applicable to line-of-business applications than others, such as ISV products, embedded systems, and so on. You may still find the discussion useful but I'm not suggesting you virtualize a standalone mobile phone application in this way!

Virtualization Is Your Friend

Making low-impact changes to complex systems used to be a nightmare. You needed a complete physical and logical replica of your system, with expensive duplicates of the hardware and software stacks. This was incredibly expensive and wasteful as, once your major changes were complete, the duplicate hardware was no longer required. I worked on one project where hardware was rented for this purpose and the delivery schedule centered on hiring and setting up this temporary kit. This is *not* an efficient way of working.

However, since the advent of utility computing and cost-effective virtualization (I know that IBM had virtualization in operating systems in the late 1970s, but these were not available to non-mainframe users until recently; it's interesting to note that this is a technology we effectively 'lost' for 25 years), this has become so easy and cheap you need a very, very good reason to not use it.

If your legacy system is not very old, or a systems team have already been through a rationalization process, then it *may* already be virtualized. This is great but you should spend some time investigating the virtualized system and upgrade this first if necessary. All the rest of the suggested steps are still valid.

50,000-Foot View of the Virtualization Process

I am lucky enough to work in environments where there is usually an expert systems team, or I can hire consultants to perform virtualization. However, I do need to know what to request. The overview below should also provide a set of pointers about what you could learn if you are not in such a fortunate position.

Terms differ between vendors, but *migration* or *conversion* are commonly used. This refers to the overall process of taking a hardware and software stack and creating a functional copy on a target virtualization platform. This usually involves:

1. Analysis of the source system
2. Creation of a target virtual machine
3. Cloning of a source system
4. Reconfiguration of the target system
5. Customization of the target system

Note that you should find out whether the process will be Physical-to-Virtual (P2V) or Virtual-to-Virtual (V2V). In a P2V process, you will be taking a system that consists of software on 'bare-metal' hardware, that is, no virtualization is already used. In a V2V process, there is already some form of virtualization used and you are just moving to a preferred platform.

It's not just servers that can be virtualized in this way but most of the network infrastructure as well. It's often possible to virtualize an entire legacy system (including all services, databases, and infrastructure) onto a single, physical host if the increase in hardware speed is enough. This allows you to have multiple copies of the system to work on – perhaps one for each member of the migration team.

Let's explore these a little more (but not too much, as these are very system- and vendor-specific).

Analysis

Your virtualization tool will provide a way to determine whether the source system is compatible and appropriate for your vendor's virtualization platform. This may involve installing a client on the source system or using a piece of software to analyze disk images, and so on. Often, it is the same client that performs this as any hot cloning.

The Creation of a Target Virtual Machine

You will need to specify the type of target virtual machine to be created. You might (and probably should) try to create one that looks as close as possible to the original source machine from the perspective of CPU cores, memory, networking, and so on, or you could boost the specification to help with performance and testing timescales.

It is usually impossible to create a target machine that looks identical to the original source machine. The new virtual machine will have 'generic' looking video and network controllers and any unusual hardware won't be emulated (some copy protection schemes will check hardware configuration to see if the software has been copied, or even use hardware dongles). If you have these issues, you should move what you can and then perform some decoupling at an earlier stage than I'd otherwise suggest.

Cloning

This is the process of copying all the data on the source machine to the target virtual machine. There are two major types of cloning available – hot and cold:

- Hot cloning – The source system is live while the clone is in progress. This allows you to perform the process on a running system and still make sure you have the latest state (very useful if you are also performing a real-time switch-over on a production system).

- Cold cloning – The source system is not running, or the state taken is not the latest. Only the stored (disk) state is copied.

The types of cloning available will depend on your virtualization platform and will also depend on your source system. Some types of cloning may not be available if your source system is very old or unusual.

Note that you may be able to create a new virtual server from some types of backup. This can be a good option if you have limited access to the legacy production systems due to paranoid system administrators.

Reconfiguration

This is the process where the virtualization tool makes necessary changes to the cloned system for it to work on the target virtualized machine. For example, making modifications to the operating system for a 'generic' network controller. Hopefully, this will be done automatically by the P2V/V2V process, but you should be aware of these changes in case of unusual errors.

Customization

You should now have a working copy of the source machine, but you might have to make some further customizations. For example, modifying the IP so it can exist on the same network.

Actions to Make a System Safe to Change

Now that I've set the scene and defined the terms, I'm going to give you my suggested set of actions to make your system safe to change.

> **Note**
>
> Be confident that you can deploy and roll back

The problem with many legacy systems is you don't really know what you have, and production is your only true copy. The temptation (especially for those from a development background) is to try to build a test system from the source code, hardware definitions, and software installations. However, you don't want to build a 'test system;' you want a copy of production that you can test. If you try and build it from scratch, after a long period of time and without the original team members, this is likely to fail.

'Therefore, you should perform these steps:

Re-Create Your Production System as a Test System

This is where virtualization technology comes in. The specifics will depend greatly on your system, but the basic process is:

1. Clone each machine in the system

2. Make the minimum changes possible for the system to run

3. Replicate network infrastructure with virtual devices

4. Connect external services or mock services

5. Block connections you don't want to occur

6. Take a backup of this state so you can recreate it at will

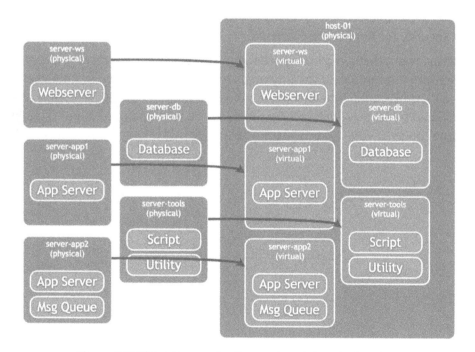

Figure 3.1: Clone the production system into a test system

This should mean you have a copy of your production system in a sandboxed environment. It thinks it's the *real* production environment but it's a system you can test and break without concern. You can recreate it to this known point at any time, which allows you to systematically measure the effects of changes you make.

Run tests and metrics for the copy

At this point, you should:

1. Gather your first set of statistics based on stress tests to see how the system performs.

2. Test the dependencies within the system to see what is required, that is, remove components you believe to be unnecessary and see if it breaks.

3. Modify undocumented configuration to find out what it does (if anything).

4. Test any 'urban legends' about the fragility of the system. (I worked on a system where the users were convinced they had to re-install the client every Monday. Completely unnecessary and I have no idea where this came from.)

5. Use and learn the system without fear of harming it.

> **Data Security for Replicated Systems**
>
> Please be aware that a replicated production system, used for testing, *must* be treated with the same data security measures as the production system itself. After all, it does contain the same data. When testing is finished, any duplicate data should be securely deleted. Do not succumb to the temptation of retaining multiple copies of your system 'just-in-case.'

Split Containers onto Their Own Virtual Machines

It's easy and cheap to create virtual machines, so, at this point, I like to split containers from shared machines onto their own. This does mean your test system isn't exactly like production but with a single service per machine you have much finer-grained control over deployment, backup, and release.

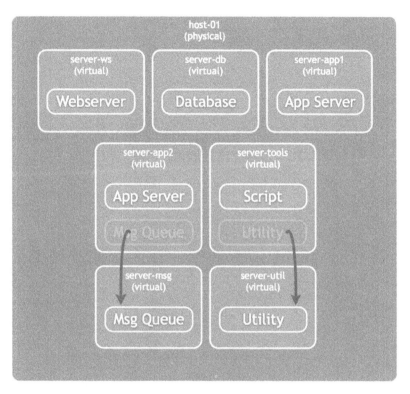

Figure 3.2: If possible, split important services onto their own virtual machines

Swap the Containers in and out

What we've done so far is replicate the system, but now we want to make sure that we can swap the individual containers in and out at will. This is so that we can make modifications to a container and deploy it within our system but swap it out if there are issues.

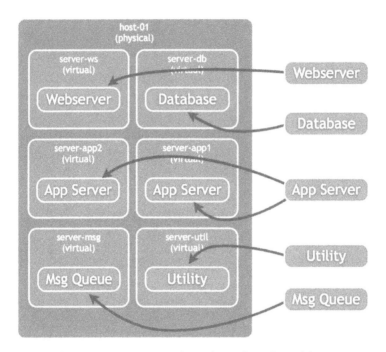

Figure 3.3: Swap Containers into Virtual Machines

The unknowns within your system and the (possible) lack of decent tests at each layer will make the failure of deployments likely. If you can swap the containers in and out easily, you can not only deploy and roll back quickly, but you also have the option of rolling forward.

> **Roll forward**
>
> You will have an intended version for release. If there is an issue with this version, you might have to roll back to the previous one. However, this is sometimes not possible; for example, if an external system, that you have no control over, has changed and you must comply with a new interface. This might mean the best option is to fix/patch the bug and roll forward to this emergency release. This is subtly different from just 'hitting the panic button' and rolling back and it's very useful to know how to quickly do this.

Once you have tried swapping containers around, you should perform some sanity checking on the system, but the likelihood is that it won't break at this stage.

Build and Deploy the Components within Containers

You should now have good control over the containers within the system. Now you can:

1. Build and deploy any components where you own the source code

2. Reinstall any third-party software

Note that you're still not trying to make functional changes; just trying to make sure you can recreate what you already have.

Logically, there should be little chance of any errors or changes in system behavior at this point, but I would suggest running a good suite of regression tests to guarantee this.

Make Some Real Changes!

You should now be confident about building your components and deploying the containers. You should also have a set of tests that you can use to check the system behavior. You are now in a great position to:

1. Try modifying some source code, build it, and deploy it

2. Upgrade third-party software in the system

3. Modify aspects of the infrastructure

You can do this with little fear, as problems should be detected easily and rectifying (roll backs) should not take long or have a large impact.

Do This in Production

I would suggest performing all the preceding steps in production as well. This may seem over the top, but you don't want a big-bang release (with a virtualized environment and large modifications) in one hit. It is much less risky to first get the methodology correct and then implement changes in small steps (you can call this 'agile' if you want), that is:

1. Virtualize your production system (preferably onto new hardware so you can keep the old system)

2. Run without changes (beyond virtualization) until you are confident it works

3. Swap containers in and out without modifications to test the code

4. Rebuild components without changes and deploy them

5. Deploy components/containers with modifications

Preparation Issues

The previous section, on making changes safely, all sounded too good to be true – and, of course, it is. There are many issues that might cause you to deviate from this plan or require a workaround. Some possible issues and suggestions for solving them follow – I encourage you to send me any that I've missed.

External Dependencies Required for Testing

This is probably the biggest issue you will face. Completely standalone systems are rare, and you are likely to have dependencies on external systems you do not control. These external systems might be anything from a simple database through to a payments system. If your system is tightly coupled, you may not even be able to start it without its external dependencies in place.

We're trying to create a test system that is as close to production as possible, so if we modify it to remove these dependencies, then the resultant test system is of limited use.

How can we solve this issue? Well, some suggestions are:

1. **Connect to an External Test System**

    If your dependency is on a commercial system, supplied by a third party, there is a good chance that they have a test/UAT service that you can connect to. You might do this by modifying your configuration or by using DNS/routing tricks to direct traffic. A virtualized infrastructure gives you great control over this.

    There are still issues with this solution, including: data being different in the test system; the test system being a different version (it might be a later version with new features); it's highly likely to have different quality/non-functional behavior.

    If your legacy system is quite old, there might not be a functioning, external test system. These are expensive to maintain and the supplier will have been tempted to turn them off. You might want to check any contracts you have to see if they are obliged to supply one or appeal to their sense of self-preservation (do they really want you testing against production?).

    If there is no external test system, then you might be able to use a disaster recovery or replicated system instead. However, there are risks associated with this if you end up with bad data in the recovery system due to your testing.

2. **Use the External System with a Test Account**

If there is no external test system, or it differs from production too much (version or data), then you might be able to connect to the real system but as a test user. The external provider will probably have to set this up (and, hopefully, duplicate any settings or data) specifically for your testing. You don't want to damage production data, so you need to duplicate anything you intend to modify.

This suggestion involves minimal modification to the legacy system being tested (user details) and the data and functionality should be realistic as it's the 'real' external system. However, you still need the help of the external system's owners and they may be reluctant to enable this in case it creates issues (substantial load, corrupt data, and so on).

> **Note**
>
> I was working on a trading system for an investment bank and had just delivered a new feature into production. I was explaining how this worked to a trader when he decided to 'test' the feature by sending himself an order (to sell a product). He had not understood the nuances of the feature (he impatiently kicked off his test before letting me finish). The order did not get sent to him but rather the guy sitting next to him – who bought the product, as it was very, very cheap. The trader who bought the product would only sell it back, to the trader who made the error, at market price (just to annoy him). There was lots of swearing. Be careful of testing in production systems.

3. **Use the External System During Downtime**

If you can't get your provider to set you up with either a test system or a test account/data, then you might consider using real users on the real system but during 'downtime'. Consider an external system, where you can place product orders, but the product orders are only processed between Monday and Friday. You could run tests against it at the weekend but cancel the orders on Sunday evening.

This requires no setup modifications on either side. However, you have to be very, very careful that you leave the external system in a good state when the downtime is over (If you create 100 orders for crates of pickled onions on the Saturday and your process to cancel the orders fails to run on the Sunday you might end up with a lot of pickled onions).

4. **Create a Dummy Service**

   Create a mock-up of the external service and route your communications to it. This has the benefit of requiring no changes to the system you are testing (you should be able to route communications to your dummy service transparently) and not relying on any external parties.

   However, if the external service is complex, you will have a lot of work in making it realistic and usable. It is likely to have different behavior and quality attributes.

   > **Note**
   >
   > Remember the feeds into and out of your system as well as external services. It can be tempting to only implement dependencies (to get your system to a working state) and forget the dependents. You need to check your outputs as well as getting good inputs.

Duplicate Test System Data Becomes Stale Quickly

Complex systems are likely to have a lot of data being constantly updated. When you create a test system, you will copy that data, but it quickly becomes stale. If you connect to external systems (using the techniques listed already), then this *raw* feed data stays up-to-date, but this is not the only way data and state are changed.

You may have more issues with manually entered data; for example, product designers entering new products. You can either perform regular operations to copy this data over (perhaps an end-of-day process) or try to create a real-time feed of this information. Feeding it in real time might be more accurate but can be hard to do for live systems. Persuading users to 'dual key' into a test system is difficult!

> **Dual Keying**
>
> This is the where a user must enter the same information into two systems. This often occurs where two systems do similar jobs, require the same information, but have no direct link; for example, a telephone salesperson entering a customer's details into a delivery system and then again into a payments system. This is a waste of time, error-prone, and very common in legacy systems. Avoid this if you can.

The modifications that operational staff perform are often forgotten but are very important. The real-time data feeds into your system can have unwanted, duplicate, or just bad information included. The operational staff will remove/clean this data, so the system operates correctly. Test systems can quickly get into a bad state if these important tasks are not performed. Remember that, on a legacy system, these tasks may be non-obvious and bordering on Voodoo.

If it's difficult to keep a test system's data in a useable condition, you may have to frequently replicate the production data over to your test system. If you've followed the advice on deployment, this should be a quick process (and you should only need to swap in specific containers).

Internal Coupling Stops You Swapping Components

Perhaps your system is so tightly coupled that you can't swap any internal containers or components. Perhaps you have a single monolithic component that does *everything*?

You might have to treat it as a single block for your deployment processes until you can start making modifications/code changes. You should aim to do this as soon as possible.

However, you should remember that you can slice and dice a system both horizontally and vertically. Most of us approach the decoupling process from a specific point of view. Those of us from a development background tend to only consider horizontal decoupling (and those from a systems background often think vertically). Try viewing your system from a different perspective and you might be able to pull more of it apart.

Changes Cause Unexpected Errors

There is a chance that moving the system can cause issues even if nothing has really changed (and even if you don't virtualize).

> **Note**
>
> I managed a team that developed software for a large-scale grid project. The computation was a Monte Carlo simulation (a Monte Carlo simulation involves a (complex) model that takes many inputs and runs a simulation multiple times while varying those inputs). You then monitor how the output varies. This is a useful technique for models where there is no simple relationship between inputs and outputs; for example, climate modelling, portfolio analysis, economic predictions, and so on) performed across hundreds of servers. The model was generating sensitivities and did this by looking for differences in outputs. When a new type of server was introduced, the outputs produced went crazy. After days of investigation, I found that there was a tiny difference in some floating-point calculations on the new machines. This meant the output could vary when the inputs didn't change. Even though the differences were tiny, it meant the sensitivity was calculated as being infinite.

Licensing and Regulation Issues Running Multiple Copies

I've already covered some of the licensing issues in the *Problems* section, but you should be especially careful of this when you're reproducing your system for testing. Your licensing might be much more restrictive than you expect – especially for databases.

You should also be careful when you are replicating data, in case you don't have the right to replicate it. There may also be privacy and copyright restrictions on doing so. If you are in doubt, speak to an expert on these legal issues.

Structural Changes for Disaster Recovery/High Availability

Until recently, you could be pretty sure that there was an almost one-to-one mapping between your deployment diagrams and your network and system diagrams. For example, your deployment node called **web server** would sit on a physical server called something like webserv-01 (with maybe a few other numbers for load balancing) and your database would sit on a physical server called **database** replicated to database-dr. The network diagram would contain a few extra boxes for items such as routers but otherwise the structure was almost identical.

Once we've rationalized and virtualized a system, this no longer holds and the assumptions we can make about the physical deployment might be wrong.

Let me give you a simple example. (Note that these diagrams are just slightly modified examples taken from http://www.gliffy.com. This is a very simple setup with some very basic load balancing/high availability.

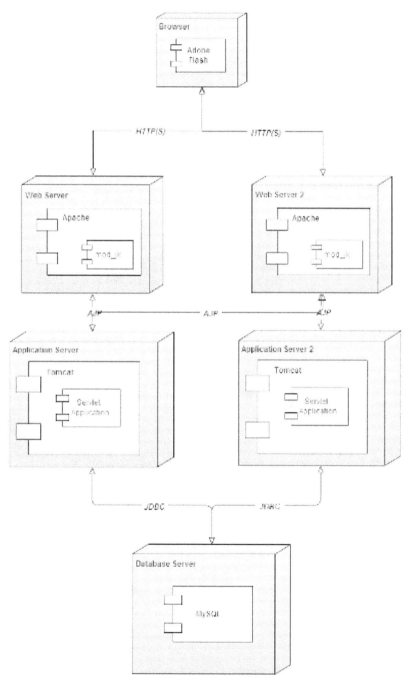

Figure 3.4: Component Deployment Diagram

We hand this over to the systems team who hand us back something that looks like this:

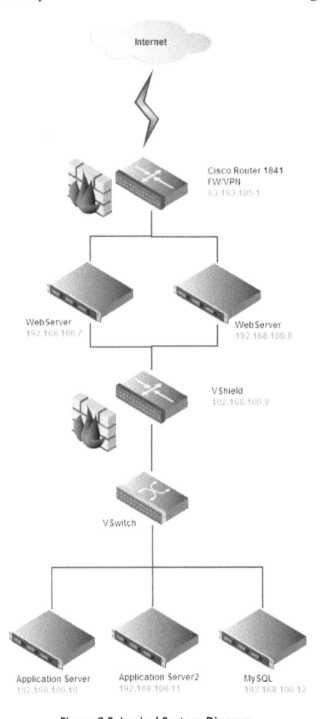

Figure 3.5: Logical System Diagram

We have a pretty much one-to-one mapping between deployment nodes and 'machines.'

Virtualization allows underused physical hardware to be shared and means that virtual machines can be shifted if they need more resources or fine-grained control over resource allocation made. So, how *might* the physical hardware look?

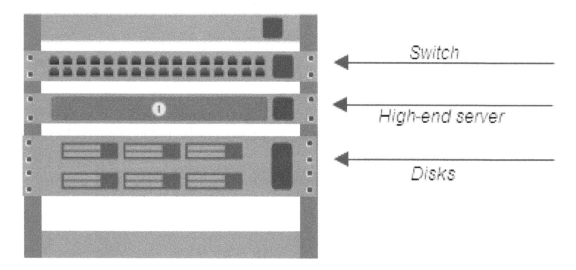

Figure 3.6: Physical Hardware Diagram

I'm being slightly facetious here, but the point is that there may be little mapping between physical hardware and virtual machines, and even your firewalls and switches may be on the same server. You can easily squeeze a couple of web servers, application servers, DBs, virtual firewalls, and switches onto a single high-end server. Externally, however, they will look like individual servers and infrastructure.

Is this an issue? Well, we've designed redundant components and we have redundant virtual machines, but our original design probably assumed redundant physical machines as well. A hardware failure could take down everything at the same time.

You must be particularly careful if you are not creating the virtual machines yourself but relying on a systems team to do this for you. The systems team may be supporting hundreds of virtual instances across dozens of application groups. The allocation of virtual machines may be automated, without any sanity checking for physical implications. The systems team may just allocate on the next available piece of hardware, so the entire application infrastructure may be on one or two machines.

In this case, we'd request that no vertical layer is deployed on a single piece of physical hardware, but we'd need a much deeper analysis on a more complex system.

Stabilization

Why Stability?

Many systems can suffer from instability and fragility. Legacy systems are particularly prone to these problems for two main reasons:

- Complex systems require maintenance, and without it, the system degrades.

- Continual modifications can add complexity and inconsistency.

These can be viewed as a kind of entropy (I know physicists are yelling that it's not really entropy but IT systems do tend toward disorder unless work is performed) and the longer a system exists, the greater the effect. If you're lucky, then your legacy system will have received all the maintenance required, but then you wouldn't have bought this book.

I would always suggest performing a stabilization phase when working with a legacy system – even if you intend to replace it. If the system is stable, then you will not be constantly pestered by user complaints.

> **Note**
>
> Stabilizing a system due to be replaced is **not** wasted effort, as it removes distractions

Not only will a stable system reduce your distractions, but users will have a better experience and it will prove your competence. Users are much more likely to share information and be helpful if they think you are competent and are trying to help them.

> **Note**
>
> I once worked on a migration project where the project manager tasked with replacing the system was also in charge of maintaining the old system. The users were reluctant to change systems (see *Common Issues* section). His response was to deliberately stop the maintenance of the old system (therefore letting it degrade) to 'encourage' the users to accept the new system. This caused frequent, critical issues (which we had to deal with) and huge friction between users and developers (the users were not stupid and knew what was going on). He would have been much better off gaining the users' trust (by improving what they had) and introducing the new system a feature at a time.

If the system is due for upgrade, migration, or modification then any stabilization will directly help you by removing potential issues before you start.

What Have We Achieved Already?

You have already performed the first stages of stabilization:

1. Understanding the system (system sketches, analysis, stakeholder analysis, and so on)

2. Hardware and container rationalization

3. Hardware modernization/performance improvements

4. System and basic functional test harnesses

5. Monitoring

These have all involved little or no changes to the system's processes, data, or logical structure. The next few stages will modify this, but users should only see non-functional (quality) attributes improving rather than functional differences.

Bad Data

We have already discussed the causes of 'rotten data' in the *Common Issues* (*Decaying Data*) section. Bad data is frequently the cause of system failure and poor performance. Common examples include:

- Circular references causing processes to hang

- Incorrectly terminating structures causing processes to crash

- Incorrect entries (Empty/Null/Wide) fields causing a dreaded `NullPointerException` or an overflow

- Large quantities of unused data slowing down queries

This cleanup falls into two main categories: removing unwanted data and correcting corrupted data.

Removing Unwanted Data

What data can be archived or removed? This is obviously application-specific but here are some broad suggestions to look out for:

- Application logs

- Inactive customers and accounts (under UK data protection rules, you are obliged to remove personal information if it is no longer required)

- Records and activities associated with removed users

- Historic transactions (remember to keep enough to comply with any data retention legislation)

- Historic 'raw' data; for example, foreign exchange values in a financial application

- Test data and accounts (often, more than you realize and easily confused with inactive accounts/data)

- Audit information (and other metadata) for data being removed

Note that the preceding data does not add value to the system and is likely to contain errors.

The first stage to removing unwanted and unneeded data should be to back it up! It's easy to make mistakes if your system lacks documentation and isn't completely understood. (I'm also assuming you'll test data removal on your test system before your production system.)

If you have virtualized your system, then you can take images for backups that are quick to restore but you should also consider exporting data in formats that can be examined later without recreating the entire system; for example, save historic customer transactions as CSV files. (Remember that backup/archived data might be subject to the same data protection rules as 'live' data. Don't just export your customer's information into CSV files and leave them on a shared drive.)

Your first attempt to remove data should be made through the application's interfaces. There may be metadata (such as counters and sizes) updated on removal and the application will hopefully do this correctly. This isn't always the case and if there are errors in the data, the application may be unable to remove it, requiring you to remove the data manually (using tools for the data store in question).

> **Note**
>
> I worked on a system that processed a large quantity of data. This data was gathered and used during the day but was not needed afterwards (it was a financial application where only end-of-day data was needed historically). It included many overnight 'clean-up' jobs that removed data that was no longer required. Parts of the system started having performance issues and it turned out that, due to a null value in the data, one of the clean-up tasks was failing silently. This led to a degradation in system performance and could only be cleared out with a database tool. After removing this one bad value, the clean-up task started working correctly again and performance improved dramatically.

I would suggest the following:

1. Attempt to remove unwanted/bad data using the application tools provided. Attempt to remove data using data store tools.

2. Shut down the application.

3. Back up the data store in question.

4. Run an analysis tool.

5. Run analysis scripts to identify data; for example, **sql/oql** selects.

6. Export the data you intend removing as an extra (and quick to examine) backup.

7. Remove/delete/archive data (depending on the technology and the options available).

8. Restart the application and make sure it is running as expected. You may only have a small window to put the data back.

By explicitly exporting the data, you can specifically re-insert it later if required. When you realize this is needed, there might be a lot more data in the system (days, weeks, or months later) and you won't be able to just roll back to the data snapshot. This is effectively creating your own archive process.

Data Cleansing

There is an important difference between data cleansing and data removal. When you cleanse data, you still retain it; you just make sure it is in the correct form, to avoid errors. Problematic data that may need cleaning includes:

- Illegal characters (such an unusual whitespace) in a comment field
- Numeric data with differing precision
- Use of **NULLs** in a database to indicate an empty string (or vice versa)
- Use of **NULLs** in a database to indicate a 'zero' value
- Overly wide fields causing data overflow (comment fields are a common violator)
- Incorrect capitalization
- Incorrect formats for phone numbers, addresses, and zip/post codes

All of the preceding can cause undesired behavior in the current systems and be rejected as data for upgrades and migrations.

Manual versus Programmatic Correction

Many errors can be present in large data stores but it's time-consuming to find them. Normally, you only become aware when an error occurs.

I would suggest doing the following when you find a data error:

1. Track down the error manually (usually by finding/tripping over a specific example)
2. Back up the data in question (this should be a simple export of the record you are looking at)
3. Manually correct the data; for example, replace a null with a valid value
4. Check that your fix has worked by rerunning your test case
5. Put the original data back (!)
6. Write a script to make the correction to this field(s) across all records
7. Run your script and check that the correction has been applied correctly

This probably sounds over-the-top, but this serves two purposes:

1. You will correct all errors of this type and not just the one you have come across. Some of these fields might not be used as often as others and could cause errors later.

2. You have a correction script that you can easily run later if these errors creep in again.

Gradually, you will build up a set of scripts you can run to cleanse your data (you will find that consultants working on common legacy systems move from job to job, taking their cleansing scripts with them and improving them as they go).

Removing Unneeded Components

As well as data, there may be entire components that are no longer used (or ever used) that can be removed. These may include items such as:

- User interfaces into the system; for example, an HTML interface when all users have a client GUI

- Components allowing communication with external systems that no longer exist; for example, defunct payment gateways

- Partially complete projects that were never used; for example, an audit system with partial coverage that has never had its data examined

If these components aren't used, then you could sensibly ask why bother removing them? The simple answer is that it gives you less to maintain and means there is less strain placed on the system infrastructure. Of course, you might turn them off and discover that they *are*, in fact, used. This is very useful information and you might need to replicate them in any replacement system.

Removing Unneeded Tasks

Many systems have tasks (or batches/processes) that run periodically to perform operations. Complex systems are likely to have many of these batch processes. These could include:

- Report generation
- Emailing reports
- Status updates
- Syncing processes
- Backups
- Data cleansing

Many of these could be required but some might not. It's common for batch processes to be set up for a one-off task and never removed.

> **Note**
>
> I worked on a system where we discovered a batch process that checked some system statistics and then contacted an SMS gateway (mobile phone text message) with a status report if certain conditions weren't met. However, no one was aware this feature existed, and the messages were sent to phones we knew nothing about (God only knows who was being spammed with this or if the phones even still worked – we were too scared to phone them and check!). We not only turned off the process but also removed an SMS proxy that had been quietly running for a very long time.

Removing unwanted processes will also reduce the load on the system and possible problems.

Process Cleansing

Process cleansing is subtly different from process removal and data cleansing. This is very applicable to batch jobs/processes that are run at specific times rather than real-time or on-demand processes; however, these are still worth examining – do you really need to run all actions whenever a specific condition is met? Often, the requirements for processes will change over the lifetime of the system; for example, they may still need to be run but perhaps not as often as when the system was originally written (or the original designers might have estimated what was required and the settings were never changed).

You might want to investigate the following:

1. What datasets are being used by the processes? Can you reduce the scope of what they are run for?

2. Check how often the processes are run. They may not need to be run as often as originally designed; for example, only once a day rather than every hour.

3. Is the process execution time still appropriate? Would it cause less impact if run earlier or later?

> **Note**
>
> I recently modified a batch process that was being run two times a day to one time. It was generating a report at the end of the business day for both the London time and New York time zones. This had been necessary in the past but was no longer required (due to the business restructuring). There were no immediate benefits, but simplifications led to a more manageable system.

Removing Unused User Options

When investigating processes and options (with the aim of removing them), you'll probably find a bunch that *are* used but rarely and by only by a small subset of users.

It's very common for the users of a system to be given more access and options than they require. The operators of the system do not know what a user will require upfront and it's easier to just enable everything rather than be pestered with access requests. You should seriously consider removing the ability for users to perform actions that they don't need.

This is good housekeeping and it's worth remembering that many serious system issues are caused by users accidentally choosing the wrong option or playing with a feature they shouldn't. Keeping an interface simple (and not cluttering it with unused options) will help the user to find what they need and reduce questions being directed to you.

Don't Underestimate a Resource Boost

It's common for a system to have gradually increasing demands made upon it and therefore gradually become less responsive. It's also possible for a system to reach a 'tipping point' and suddenly become completely unusable. There are many books about non-IT complex systems that reach tipping points and the effects that occur; these range from population (usually a crash) to financial markets (also usually a crash). If you want an interesting popular science book on the subject, I recommend *The Tipping Point: How Little Things Can Make a Big Difference* (ISBN 0-316-31696-2) *by Malcolm Gladwell.*

> **The Overnight Batch**
>
> It's a common misconception amongst IT managers that 'overnight' is an infinitely long period of time, during which a computer can complete any task given to it. When a task is run overnight, it can gradually take longer to complete but won't be noticed – as long as it has completed by morning. Of course, one morning you'll arrive to find it *hasn't* completed...

Often, the simplest and cheapest way to make a system more useable is to add some resources. If you've virtualized your system, then assigning extra RAM or CPU might be easy, or you might need to physically add more resources to the hardware.

Don't worry about this being a wasted investment (particularly a concern if you ultimately intend replacing the system); it can quickly pay for itself in saved time, and with modern virtualization platforms you can easily re-provision afterward.

Lastly, remember that resources aren't just CPU and memory; IO for communications and storage speed also have an impact. Consider upgrading your network connections (you might have an old network card installed that doesn't run at the full speed of a modern network) or replacing your disks with faster ones. If your CPU isn't being utilized, then maybe the data just isn't getting there fast enough.

Re-Examine your Application Settings

There are a variety of application and platform settings that can be used for tuning. Examples include:

1. Database cache settings

2. Database process settings

3. Transactional guarantees

4. JVM/CLR memory limits (moving to a 64-bit JVM allows you to use more memory)

5. JVM/CLR garbage collection settings or methods

6. Queuing strategies

The system might have been well-tuned when it was first installed but if the use has changed, you might need to re-examine these settings. What is available will depend on your system specifics but remember to examine all the containers *and* look up and down the stack for tuning opportunities.

Optimize the Users' Actions

The users might not be using the system in the most efficient way (refer to sections *No Documentation*, *Lost Knowledge*, and *Hidden Knowledge*). While you are investigating the system, you might discover better ways of working that you can inform the users about.

You can also spend some time creating macros or templates within the application to streamline repetitive work. If you observe how users use the system, then you might be able to make some simple changes that make a big difference.

> **Note**
>
> I once worked on a project that had a heavyweight GUI client with complex, nested menu options. One common feature was hidden deep within these menus and users wasted time every day simply selecting it (some of the sub-menus needed careful mouse-pointer control or the whole tree would collapse). I configured the GUI so a shortcut to this feature was added on the main toolbar. This was probably the most appreciated work I've ever done, and it took 5 minutes.

Leaving a Good Legacy

Why Bother Leaving a Good Legacy?

You may be asking yourself a question such as:

"If I spend time and money worrying about what happens in the distant future, might I reduce my chances of success in the short term?"

This is, of course, true. If a project fails in the short term, then it will never *become* a legacy system. Therefore, you need to be careful in balancing maintainability in the long term with success in the short term – but don't go to extremes at either end.

However, there are many good reasons to worry about leaving a good legacy:

1. You may be in your current job a lot longer than you anticipate.

2. You may be asked to return to this job/project for a new phase.

3. We live in a small world (particularly if you work in a specific business sector within IT) and your reputation can suffer if you leave behind an unmaintainable mess.

4. Making sure you create a quality system for the long term is also likely to make sure it is a quality system overall. Maintainable systems tend to be of higher quality and easier to operate.

So far, I've discussed problems, strategies, and techniques for dealing with a legacy system that you inherit. Much of this assumes there are negative issues to be faced. How can you avoid creating these issues yourself when working on a new system?

Documentation

The documentation you produce doesn't have to be heavyweight or comprehensive (and it's less likely to be maintained and useful if it is). What your stakeholders require is the basic information that is not obvious from either the code, configuration, or the system's visible interfaces.

This information doesn't have to be written in a traditional document but can be stored in any form that is convenient. This can include documents, internal websites, wikis, code comments, configuration comments, application help, naming conventions, and so on. However, you should be careful to make sure it is stored in a form that is unlikely to be deleted or lost. If the information is co-located with the product (for example, application help files are usually stored with binaries), then it will be available for as long as the product is used. If this is not the case (for example, an internal website created by a development team), it is highly likely to be removed. This is especially true if keeping it incurs a cost. Consider keeping a copy of any documentation in the source control system along with the code so that there is a single place/artefact for the system. Some information you should consider documenting includes build and deploy Instructions, access information, location information, test scripts, asset register, startup and shutdown instructions, business continuity and disaster recovery, backup plans, software guidebook and the big picture.

Build and Deploy Instructions

I have seen systems where the source code is correctly stored in a source control system but without any meta-information about how to build and deploy. Even if the build system you use (such as **Ant**) has a default target for a build, you should explicitly state that this is the correct target. If you are using a different target, then you *must* state this, or maintainers are almost certain to use the default incorrectly.

You should also record the latest branch/label used and include a description of the labelling/branching rules. Please state which are release or development branches and if labels indicate stable versions. It is very difficult to interpret this meaning later.

Once the artefacts from the build have been generated, the future maintainers will still need to know what to do with them. Most full deployments (that is, a new system) are more complicated than is necessary and the creators of the system may perform important actions without ever documenting them. If the knowledge of these steps is lost, then the system might be impossible to re-install. It is good practice to get new team members (or someone from outside the development team) to try to install and configure your system from scratch. If they cannot do so, without help from a current team member, then your instructions are insufficient.

I would suggest storing this documentation as simple text files within the source control system itself. Simply place it in the root directory and give it an obvious name, such as `build_instructions.txt`.

Access Information

Access information is the information required to gain administrative (or another level of) access to any subcomponent of the system. This includes usernames, passwords, schema IDs, and so on. You need to store both the authentication information (usernames with corresponding passwords), how to input them, and the services they access. For example, in a database, you may need username, password, relevant schema-name, database instance, and database location.

This information may be sensitive and requires analysis on how to store it safely but you *must* make sure that it is available.

If you have many types of users, with differing levels or combinations of authorization, then describe these. Specify the reasoning for this complexity so that security is not compromised by a new user being given unsuitable access.

Location Information

Along with *how* to access data, you need to describe where it can be found. Although full data schemas can be useful, it is also helpful to separately describe where specific bits of important information reside. For example, if your system uses a dependency injection framework with multiple XML files, it is helpful to explicitly identify the root file that links the others together. Types of data you could give locations for include:

- Configuration files
- Configuration items
- Initialization parameters
- Links
- Binaries

The locations you list could include:

- Tables, columns, row IDs in a database
- Locations on a file system
- The latest branch in source control
- Locations within archive files such as JAR files, ZIP files, and so on
- Network locations
- Machine locations
- FTP destinations/sources

Third-party systems often ship with large amounts of configuration or multiple methods of execution that are irrelevant to your specific installation. Their documentation will cover everything. This can confuse future maintainers so you should reference what within the third-party system's documentation is applicable.

Test Scripts

If a QA team manually tests your system by following **test scripts** (by 'scripts' I am referring to a set of manual instructions for a person to follow rather than an executable script for a machine), then please keep these! All too often, they are viewed as transient information and are not stored; however, they are an invaluable source for showing how the system is intended to work.

I would also suggest storing these in the source control system, as each version of the system is likely to have an associated set of test scripts. By storing them in the source control system, the relevant test scripts are with each released/labelled version.

Asset Register

An **asset register** is an accounting document that lists all the assets that an organization owns. If you are creating or maintaining an IT system, then it is beneficial to make sure that all the hardware and software assets are recorded in this asset register.

If hardware assets are registered, they are more likely to have maintenance contracts paid and keep associated support. They are also less likely to be accidentally decommissioned, as the asset register will not only list the assets but also the departments and projects they are used in. This is of concern if a system has more than one owner – if one owner stops using the system, the organization may believe that it is now redundant.

Software assets may also benefit from ongoing maintenance and avoiding decommissioning, but they are also more likely to have complex licensing contracts. The asset register should reference the license agreements for software, which will be useful if the organization needs to know if future changes of use are legally allowed.

Startup and Shutdown Instructions

Clear instructions on bringing the system up and down are often missing from legacy applications. In particular, instructions for gracefully shutting down the system in whatever state it is in (including errors states) is often poor or absent.

A well-designed system shouldn't be kept in order for starting, but if this is important, then please explicitly state this. When first deployed, the operation staff may be aware of this knowledge, but it can easily be lost with staff changes.

Shutting down the system gracefully when it is in an error state may happen infrequently, so a detailed, stepwise description can be invaluable. Again, if the order is important, then please explicitly state this.

Business Continuity and Disaster Recovery

As well as documenting how to start and stop the system under normal (or simple failure) conditions, you should also create instructions for dealing with major disaster scenarios. This is a complex topic (I include some links in the appendix) but ultimately involves a set of instructions for completely recovering a system. This differs from a failure, in that the entire system (potentially all the components but at least enough to render it inoperable) stops working and cannot be started – a complete recovery at different physical premises may be required. Many software architects don't consider these scenarios, but they can include:

1. Power surge leading to a datacenter failure

2. Flooding leading to the damage of machines located in a building's basement

3. Communications failure leading to loss of connectivity

4. A disgruntled staff member deliberately corrupting a system

5. A staff member accidentally shutting down the system

6. Terrorist action leading to the loss of a building with re-entry impossible

These are usually classified into either natural or man-made disasters. Importantly, these are likely to cause outright system failures and require some manual intervention – the system will not automatically recover. Legacy systems may have been running perfectly well for many years when a physical disaster strikes. There may be no one left in the organization who remembers how the installation was first performed, so an organization should have a **Disaster Recovery** (**DR**) plan for operational staff to follow if this occurs.

A disaster recovery plan should consider a range of scenarios and give very clear and precise instructions on what to do for each of them. In the event of a disaster, the staff members are likely to be stressed and not thinking as clearly as they would otherwise. Keep any steps required simple and don't worry about stating the obvious or being patronizing – remember that the staff executing the plan may not be the usual maintainers of the system.

The aims and actions of any recovery will depend on the scenario that occurs. Therefore, the scenarios listed should each refer to a strategy that contains some actions.

Before any strategy is executed, you need to be able to detect that the event has occurred. This may sound obvious, but a common mistake is to have insufficient monitoring in place to detect events. Once detected, there needs to be comprehensive notification in place so that all systems and people are aware that actions are now required.

For each strategy, there must be an aim for the actions. For example, do you wish to try to bring up a complete system with all data (no data loss) or do you just need something up and running? Perhaps missing data can be imported later or maybe some permanent data loss can be tolerated? Does the recovered system have to provide full functionality or is an emergency subset enough?

This is hugely dependent on the problem domain and scenario, but the key metrics are recovery point objectives (RPO) and recovery time objectives (RTO), along with level of service. Your RPO and RTO are key non-functional (quality) requirements and should be listed in your software architecture document. These metrics should influence your replication, backup strategies, and necessary actions.

The disaster recovery plans for IT systems are a subset of the broader 'business continuity' plans (BCP) that an organization should have. This covers all the aspects of keeping an organization running in the event of a disaster. BCP plans also include manual processes, staff coverage, building access, and so on. You need to make sure that the IT disaster recovery plan fits into the business continuity plan and you state the dependencies between them.

There are a range of official standards covering **Business Continuity Planning**, such as ISO22301, ISO22313, and ISO27031. Depending on your business and location, you might have a legal obligation to comply with these or other local standards. I would strongly recommend that you investigate whether your organization needs to be compliant – if you fail to do so, then there could be legal consequences.

Backup Plans

All systems need to have their data backed up. This is system dependent and varies greatly, so it must be documented. Even if the processes are largely automatic (such as automated off-site copying), operational staff will need to know how they work in case of any failure in the process. It is common for legacy systems to have their backup processes fail silently, which is only discovered in a disaster scenario – that is obviously too late. There is no point having a complex disaster recovery plan if there is no data to recover!

Software Guidebook and the Big Picture

All the preceding documentation can be included in a software guidebook. This should also include a section giving an overview of the system as a whole and its purpose. It is much easier to maintain a system if people know how it fits into the real world. Simon Brown's book *Software Architecture for Developers*–https://leanpub.com/software-architecture-for-developers gives an overview on writing one.

You should make sure that you include some architecture diagrams to show the context that the system operates in and the major internal and external systems that it communicates with.

A containers diagram will also clearly show where the major components run from and their communication protocols. It is likely that the startup, shutdown, backup, and disaster recovery documents will be broken down by containers as well, so you should link this diagram and the document sections that apply to each element.

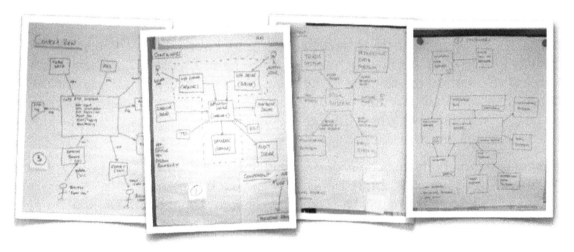

Figure 3.7: Architecture diagrams don't have to be too detailed or official – just make sure they are useful and relevant.

Tests

It is true that many legacy systems contain few tests. This may be because they were never written, have been lost, or the development team relied on a QA team performing manual tests (these manual tests probably had test scripts, which hopefully have been retained).

Tests are very useful – not only for verifying the system but also for showing its intended use. **Test-Driven Development** is currently very popular, and this helpfully leads to a high number of tests being stored with source code. However, this can result in an obsessive focus on unit testing with many project teams neglecting other forms of testing.

Unit tests are fantastic for functional testing at the class and component level, but they are less useful for any non-functional or integration testing. (They are also of no use for user-experience testing.) You should consider major use case scenarios and produce tests that exercise those conditions (Philippe Kruchten describes how to use scenarios to illustrate and evaluate IT architectures in his **4+1 View Model**. It describes sequences of actions between objects and processes. See *Kruchten, Philippe (1995, November). Architectural Blueprints - The 4+1 View Model of Software Architecture*–http://www.cs.ubc.ca/~gregor/teaching/papers/4+1view-architecture.pdf.

Some basic descriptions for tests you should consider including with your system in order to leave good artefacts for those maintaining it follow. The descriptions are brief and if you are not already familiar with them, I suggest further reading.

System and Smoke Tests

System testing is conducted on a complete, integrated system to evaluate compliance with its specified requirements. System testing falls within the scope of black-box testing and should require no knowledge of the inner design of the code or logic (this is the definition used in the IEEE Standard Computer Dictionary: A Compilation of IEEE Standard Computer Glossaries; IEEE; New York, NY.; 1990).

In other words, you should produce some test scripts that describe how to perform end-to-end tests on an installed and integrated system. You may include tests for most of the required features or (if this is too heavyweight) you could just include a set of tests that touch every part of the system, that is, test a set of spikes through the system that show all components are connected and initialized. This reduced test set is often referred to as **smoke testing** (or build verification testing).

Well-defined smoke tests are incredibly useful for staff working with legacy systems. They may not have deep knowledge of the system but might need to make small changes to the underlying infrastructure on a regular basis. A lightweight set of smoke tests allows them to perform maintenance actions and make sure they haven't created a major issue by mistake. Smoke tests can also be used to verify successful business continuity execution. Put a set of smoke tests in your software guidebook!

Integration

Integration testing is closely related to system testing (it is black-box testing of functionality) but it differs in that components are combined and tested as a logical group, that is, not the entire system but a testable subset. This allows a building-block approach, where these groups can be combined to make the full system.

This has the benefit of allowing the system to be broken down for analysis and any errors being associated with a logical grouping rather than only knowing that the system as a whole has an error. Therefore, the actual cause of errors can be pinpointed more accurately and quickly. It helps any general, operational staff to locate an appropriate, specialized team.

There are a variety of patterns for integration testing and these define how to break a system down into testable logical groupings. Some extra software infrastructure may be needed to support these patterns. I will not go into further details here, but please see the bibliography for suggested reading.

Non-functional and Quality Attributes

Functional testing is covered by system, integration and acceptance tests but non-functional or quality requirements may require some specific, extra testing. You need to define exactly what the quality requirements are *and* how to test them. It is very easy to give a vague requirement (such as "it must be fast") and not define where and how this applies. Tests you should consider including with the system, integration, and acceptance tests should cover:

- Performance (throughput and latency)
- Jitter/variance of latency
- Soak/endurance/stress
- Load
- Destruction

There are many other non-functional tests that I could list (such as internationalization testing and localization testing) but these are helpful to the software development process rather than as useful artefacts for future support. Defining the quality attributes and how to test them not only allows future maintenance teams to avoid regression from modifications but also to identify whether the system is degrading over time. Performance degradation over time is a common issue with legacy systems.

Design Considerations

Technology Consistency and Scope

Developers love to use technologies and it can be tempting to try to solve every problem by introducing a new tool. If this is not constrained, then there can be a huge number of infrastructure elements used. This creates two potential problems for later maintenance: consistency and support scope.

Without consistency, each part of the system needs to be supported differently and this in turn requires different skills. Consider the difficulty in supporting a system that uses several programming languages, middleware, and multiple databases compared to one that uses a single language, with well-known middleware and a single database. A stack with a minimal technology set requires less time for new staff members to become familiar with it.

Please consider whether any technology you introduce is necessary and whether it duplicates functionality. If there is duplication, then either don't use it or completely replace the other technology it duplicates. Try to keep a single tool for a single job.

New Technologies

Also consider the issues caused by using promising, new technologies that are dropped by their creator (or the company producing it going bankrupt) with no support. The computer industry is littered with examples of abandoned software tools. Systems using these tools have unsupported components and technologies. This can lead to problems finding skilled workers and difficult migration efforts later.

Be careful using technologies that may not be supported in the long term. This is difficult (just ask any stock analyst specializing in technology companies) but product owners with a long-term product roadmap tend to better support legacy products.

New technologies often have unforeseen problems that no one has discovered or fixed yet. Examples may include performance issues at scale or bugs caused by external events such as date rollovers.

One way of managing this is to make use of open source software that you can support yourself. However, you must consider whether you want to maintain your main system and all the tools it relies upon.

Fail-Safe

When designing a system, consider how it could fail and what would happen if it did. One of the most misunderstood engineering terms is 'fail-safe'. Most people from a non-engineering background (including many software developers) believe it means something won't fail.

A **fail-safe** device/system is expected to eventually fail but when it does it will be in a safe way. Classic examples include brakes on trains that engage when they fail and ratchet mechanisms in lifts/elevators so they can't drop if the cable breaks. Well-engineered physical devices state their **Mean Time Between Failure** (**MTBF**) and define how they can fail and what happens when they do. A well-maintained physical device may never fail in its lifetime, but you know what will happen if it does.

As there is no physical wear and tear on a software system, the concept of MTBF is arguably not applicable (although it is on the hardware side). However, software systems can and do fail all the time, so perhaps it's surprising that many software systems don't cope with failure very well or have defined actions when they fail. For example, the following may happen:

1. Underlying hardware failure; network and external disks are the ones I encounter most.

2. External system failure. Even if your system is perfect, the external systems you rely on might feed you garbage.

3. User error. If you create an idiot-proof system, then I guarantee they will employ a better idiot.

It's tempting to try to correct a failure situation and keep on running, but this can lead to a system getting into an unknown state and creating more issues; for example:

- The network is not responding but you keep on processing inputs and queuing outputs, hoping it comes back. Your caches and disks fill up, affecting other systems. Eventually, it does come back online, and your system stops responding as it processes hours' worth of stale data.

- An external data provider starts sending blanks in a numeric field. A developer had previously decided to 'interpret' empty as a zero (whereas it was missing data) and this fed through a bank's pricing system and was forwarded onto another system, which then tried to execute buys (as they were obviously a bargain at zero!).

- In finance applications, we worry about 'fat fingers,' where a trader hits the wrong keys and buys 12 million rather than 1 million.

All the preceding examples are real ones I have come across and the systems should have put themselves into a known, safe state instead. For example:

- Put limits on anything you do for recovery situations; for example, retry only three times, put a time limit on caches, and so on. Don't continually perform an action that isn't working.

- Don't make generic assumptions about correcting data across a system. If it's not a good input, then fail that input, as you have no idea what it really means and you are hiding the error. Note that I'm not suggesting the entire system should be suspended but the transactions that has error should be suspended and reported upon.

- User inputs are often sanity checked but 'are you sure?' dialogs are automatically clicked (without reading them) or the 'never show this again' checkbox is selected. Ultimately, there is only so much you can do to save users from themselves, but you might want to save an audit from users' decisions.

It's important to not just put the system (or transaction) into a safe state but to also inform those that can resolve the situation. Developers often log errors and think nothing more about it. It's amazing to use a reporting tool on a legacy system and extract all the worrying messages. Would it be more appropriate to send an email, pager message, text message, or change a dashboard status, and so on?

You need to design error reporting and monitoring services up-front and define how the operators should be kept informed. The operators need to be able to resolve issues speedily and safely.

Example Legacy Scenario

A Very Brief Overview

The salient points:

1. You work for a furniture company

2. You get a promotion!

3. You are now responsible for the warehouse inventory system...

4. No one has touched it for a while

5. It was written in 2003

6. It's a 3-tier architecture

7. It basically works, although people moan about it.

8. What do you do?

Why this example?

Most organizations are *not* primarily focused on technology! One of the reasons I have picked a furniture company is that any IT systems they have will support the main business rather than being the revenue driver. This is true for most IT systems in most organizations – web companies selling online services are not representative.

You have found yourself responsible for this system and discovered that no one has *developed* (or performed maintenance) programming on this system for a while. However, the system is used and performs an important and core (supporting) function within the business – no one gets their furniture if the warehouse is not operating correctly.

How should you respond to the situation you find yourself in and what problems are you likely to face? What are the strategies you should employ and how should you execute them?

It's very difficult to get a good impression of what this system does, who uses it, and how they use it from the description above (or any text-only description). Hopefully, the diagrams and sketching techniques used in the planning chapters demonstrated how useful they can be.

B

Legacy Project Questions

Why a List?

Much of the advice I have offered is spread across several chapters and involves the use of anecdotes.

Therefore, I have built a list of questions that can be used when someone takes ownership of a legacy project, to help identify issues and problem areas. You should try to get these answered as-soon-as-possible. Take the list of questions with you when talking to those who have experience of the system.

In order to de-risk your system you should be especially careful of anything that is unavailable or will take a long time (such as access to hardware).

The list is broken down into sections and some (such as source code) may, or may not, be applicable to your project. These lists are not comprehensive, and you should add items relevant to your industry and technology.

Section	Question	Notes
System Overview	How long ago was the system originally developed?	Try to get a handle on the real history
	When was it last officially released?	
	When was it last modified?	This may or may not be part of an official release. Careful!
	How frequently has it been modified?	Try to find out if it is active.
	Who performed any modification?	You need to find out who has knowledge of changes and motivation
Physical Hardware	Where are the systems physically located?	This may be much harder to find out than you realize
	Do you/someone have physical access to the systems?	
	What are the timescales required to get physical access?	There may be long approval times to build into your plans
	Is any hardware still in warranty?	Does the organization that made it still exist?
	Is any hardware still supported?	

Section	Question	Notes
	Can you buy identical replacement hardware?	If not, you may need to test similar physical/virtual hardware or buy second hand
	Are any test servers available?	These are likely to have been retired and may need recreating
Documentation	Is there any documentation?	Official documentation and unofficial, personal notes
	Where is any documentation stored?	
	Are there multiple versions of the documentation?	Careful as it may differ from the systems deployed
	What format is the documentation in?	Is it still usable?
	Do you have release notes/process?	These are often lost for old systems
External Services	What external services are required to run?	Add these to a context diagram
	What external systems rely on your target system?	i.e. will potentially break if it was decommissioned
	Are external test services available?	It'll be hard to test otherwise, and these may no longer be available
	Is there any documentation for communication protocols for external systems?	

Section	Question	Notes
	What are the contact details for the owners of reliant, external systems?	
	Do you have contact details for the administrators of external systems?	
	What are the main data feeds into the system?	
	What are the main data feeds out of the system?	
	What data verification is performed on data feeds?	
Internal Services	What internal servers/ services are required for the system to run	For example, a centralized database. Add these to your context diagram
	Are the administrator credentials known for internal services	You may need to make modifications.
Data	What data quality checks are periodically performed on the system's data?	
	What archiving or data deletion is periodically performed on the system's data?	
	Do you own all the data in the system?	Some of the reference/static data may not be yours

Section	Question	Notes
	Is any of the data covered by licensing restrictions?	Common in financial services
Sourcecode	Is the sourcecode available?	
	Is the history of the code available in source control?	
	Do you have the access (usernames/passwords) for any source control?	
	Are any of the original developers/installers still employed?	Including those that have performed configuration
	Is the build configuration and test code still available?	
	Is a build server still available?	
	Does the code build?	
	What was the last release version?	Can you identify in the source control?
Configuration	Do you know all the software in the stack?	And the exact version of that software.
	Do you have access to the installation media for third party software?	This will be important if you need to rebuild a stack.
	Is any third-party software still supported?	Does the organization that made it still exist?

Section	Question	Notes
	Can you get copies of licenses for third party elements of the system?	You may not be allowed to make some required changes and legal advice is time consuming.
Users	Can you get a list of all current users?	Use this to start a User's/ Concerns table
	Do you have physical locations and contact details for all users?	You may need their 'buy-in'. Contact sooner rather than later.
	What are the systems roles for all the users?	
	What functionality of the system do each of the user's access?	So, you can ask the right user about the right area
Known Issues	Are there any obvious bugs the users know about?	
	Are there any known performance issues?	
	Does the system ever crash or become unresponsive?	And find out how this is resolved.
	Are parts of the system replaced by other, newer systems?	There may be chunks of the system you don't need to support
	Is the system covered by any regulation?	For example, financial, privacy, data protection, safety and so on
	Is the system covered by and business continuity or disaster recovery plans?	

Section	Question	Notes
	Is there any high availability planning for the system?	
	Does the system have uptime or other performance guarantees?	
	Who is responsible for the costs of any modifications to the system?	

References and Bibliography

This is closer to a 'suggested reading list' than a formal reference section or bibliography. I have enjoyed and found the following books, videos and references useful and I hope they are good suggestions for further reading. If there are any, I've missed out then please contact me!

- "Software Architecture for Developers" by Simon Brown

  https://leanpub.com/software-architecture-for-developers

- "A Team, A System, Some Legacy ... and You" by Eoin Woods

  http://www.infoq.com/presentations/Legacy-Systems

- "Software Systems Architecture" by Eoin Woods and Nick Rozanski

  http://www.amazon.co.uk/Software-Systems-Architecture-Stakeholders-Perspectives/dp/032171833X

- "Averting Tragedy on the Architectural Commons" by Robert Smallshire

  http://vimeo.com/49377271

- "Working Effectively with Legacy Code" by Michael Feathers.

  http://www.amazon.com/Working-Effectively-Legacy-Michael-Feathers/dp/0131177052/

- "Refactoring: Improving the Design of Existing Code" by Martin Fowler, Kent Beck, John Brant, William Opdyke, Don Roberts

  http://www.amazon.com/Refactoring-Improving-Design-Existing-Code/dp/0201485672

- Information Security and Systems Architecture

  Security Quality Requirements Engineering

  http://resources.sei.cmu.edu/asset_files/TechnicalReport/2005_005_001_14594.pdf

  A *Taxonomy of Operational Cyber Security Risks*

  http://resources.sei.cmu.edu/asset_-files/TechnicalNote/2014_004_001_91026.pdf

  The xkcd advice on passwords http://xkcd.com/936/

- VMWare documentation on virtualization including

  http://www.vmware.com/pdf/virtualization.pdf

  http://www.vmware.com/files/pdf/techpaper/Enterprise-Java-Applications-on-VMware-Best-Practices-Guide.pdf

  http://www.vmware.com/files/pdf/virtual_networking_concepts.pdf

  http://www.vmware.com/pdf/Perf_Best_Practices_vSphere5.0.pdf

- *The Tipping Point: How Little Things Can Make a Big Difference* (ISBN 0-316-31696-2) by Malcolm Gladwell.

- *Testing Object-Oriented Systems: Models, Patterns, and Tools* (ISBN 0-201-80938-9) by Robert Binder

- ISO/IEC/IEEE 42010 Information and Reference

  http://www.iso-architecture.org/ieee-1471/

- Institutional memory and reverse smuggling by 'An engineer', 2011-12-04

  http://wrttn.in/04af1a

- Expert Judgement on Markers to Deter Inadvertent Human Intrusion http://prod.sandia.gov/techlib/accesscontrol.cgi/1992/921382.pdf

Index

About

All major keywords used in this book are captured alphabetically in this section. Each one is accompanied by the page number of where they appear.

www.ingramcontent.com/pod-product-compliance
Lightning Source LLC
Chambersburg PA
CBHW080535060326
40690CB00022B/5138